Process Design, Integration, and Intensification

Process Design, Integration, and Intensification

Special Issue Editors

Mahmoud El-Halwagi
Dominic C. Y. Foo

MDPI • Basel • Beijing • Wuhan • Barcelona • Belgrade

MDPI

Special Issue Editors

Mahmoud El-Halwagi
Texas A&M University
USA

Dominic C. Y. Foo
University of Nottingham
Malaysia

Editorial Office
MDPI
St. Alban-Anlage 66
4052 Basel, Switzerland

This is a reprint of articles from the Special Issue published online in the open access journal *Processes* (ISSN 2227-9717) from 2018 to 2019 (available at: https://www.mdpi.com/journal/processes/special_issues/process_integration).

For citation purposes, cite each article independently as indicated on the article page online and as indicated below:

LastName, A.A.; LastName, B.B.; LastName, C.C. Article Title. *Journal Name* **Year**, *Article Number*, Page Range.

ISBN 978-3-03897-982-1 (Pbk)
ISBN 978-3-03897-983-8 (PDF)

Cover image courtesy of Melwynn Leong.

Contents

About the Special Issue Editors vii

Dominic C. Y. Foo and Mahmoud El-Halwagi
Special Issue on "Process Design, Integration, and Intensification"
Reprinted from: *Processes* **2019**, *7*, 194, doi:10.3390/pr7040194 . . . . **1**

**Doris Oke, Thokozani Majozi, Rajib Mukherjee, Debalina Sengupta and
Mahmoud M. El-Halwagi**
Simultaneous Energy and Water Optimisation in Shale Exploration
Reprinted from: *Processes* **2018**, *6*, 86, doi:10.3390/pr6070086 . . . . **3**

**Steve Z. Y. Foong, Viknesh Andiappan, Raymond R. Tan, Dominic C. Y. Foo and
Denny K. S. Ng**
Hybrid Approach for Optimisation and Analysis of Palm Oil Mill
Reprinted from: *Processes* **2019**, *7*, 100, doi:10.3390/pr7020100 . . . . **26**

Mohammed Alghamdi, Faissal Abdel-Hady, A. K. Mazher and Abdulrahim Alzahrani
Integration of Process Modeling, Design, and Optimization with an Experimental Study of
a Solar-Driven Humidification and Dehumidification Desalination System
Reprinted from: *Processes* **2018**, *6*, 163, doi:10.3390/pr6090163 . . . . **53**

Takehiro Yamaki, Keigo Matsuda, Duangkamol Na-Ranong and Hideyuki Matsumoto
Special Issue on "Process Design, Integration, and Intensification"
Reprinted from: *Processes* **2018**, *6*, 241, doi:10.3390/pr6120241 . . . . **88**

Shu Yang, San Kiang, Parham Farzan and Marianthi Ierapetritou
Optimization of Reaction Selectivity Using CFD-Based Compartmental Modeling and
Surrogate-Based Optimization
Reprinted from: *Processes* **2019**, *7*, 9, doi:10.3390/pr7010009 . . . . **99**

Augustine O. Ifelebuegu, Habibath T. Salauh, Yihuai Zhang and Daniel E. Lynch
Adsorptive Properties of Poly(1-methylpyrrol- 2-ylsquaraine) Particles for the Removal of
Endocrine-Disrupting Chemicals from Aqueous Solutions: Batch and Fixed-Bed Column
Studies
Reprinted from: *Processes* **2018**, *6*, 155, doi:10.3390/pr6090155 . . . . **119**

Muhammad Abdullah and John Anthony Rossiter
Input Shaping Predictive Functional Control for Different Types of Challenging Dynamics
Processes
Reprinted from: *Processes* **2018**, *6*, 118, doi:10.3390/pr6080118 . . . . **133**

**Lukas Uhlenbrock, Maximilian Sixt, Martin Tegtmeier, Hartwig Schulz, Hansjörg Hagels,
Reinhard Ditz and Jochen Strube**
Natural Products Extraction of the Future—Sustainable Manufacturing Solutions for Societal
Needs
Reprinted from: *Processes* **2018**, *6*, 177, doi:10.3390/pr6100177 . . . . **151**

About the Special Issue Editors

Mahmoud El-Halwagi is a Professor and Holder of the Bryan Research and Engineering Chair at the Artie McFerrin Department of Chemical Engineering, Texas A&M University, and is the Managing Director of the Texas A&M Engineering Experiment Station's Gas and Fuel Research Center. Dr. El-Halwagi's main areas of expertise are process integration, synthesis, design, operation, and optimization. Specifically, Dr. El-Halwagi's research focuses on sustainable design. In addition to the theoretical foundations he helped lay down in these areas, he has been active in education, technology transfer, and industrial applications. He has served as a consultant to a wide variety of chemical, petrochemical, petroleum, gas processing, pharmaceutical, and metal finishing industries. He is the co-author of more than 400 papers and book chapters, the co-editor of six books, and the author of three textbooks. Dr. El-Halwagi is the recipient of several awards, including the American Institute of Chemical Engineers Sustainable Engineering Forum (AIChE SEF) Research Excellence Award, the National Science Foundation's National Young Investigator Award, the Lockheed Martin Excellence in Engineering Teaching Award, the Celanese Excellence in Teaching Award, and the Fluor Distinguished Teaching Award. Dr. El-Halwagi received his Ph.D. in Chemical Engineering from the University of California, Los Angeles and his M.S. and B.S. from Cairo University.

Dominic C. Y. Foo is a Professor of Process Design and Integration at the University of Nottingham Malaysia Campus, and is the Founding Director for the Centre of Excellence for Green Technologies. He is a Fellow of the Institution of Chemical Engineers (IChemE), a Chartered Engineer with the UK Engineering Council, a Professional Engineer with the Board of Engineers Malaysia (BEM), as well as the past Chair for the Chemical Engineering Technical Division of the Institution of Engineers Malaysia (IEM). He is a world-leading researcher in process integration for resource conservation. He establishes international collaboration with researchers from various countries in Asia, Europe, America, and Africa. Professor Foo is an active author, with five books and more than 140 journal papers, and has made more than 200 conference presentations, with more than 30 keynote/plenary speeches. He served on the International Scientific Committees for many important international conferences (CHISA/PRES, FOCAPD, ESCAPE, PSE, etc.). Professor Foo is the Editor-in-Chief for Process Integration and Optimization for Sustainability (Springer Nature); Subject Editor for Trans IChemE Part B (Process Safety and Environmental Protection, Elsevier); and is an Editorial Board Member for Water Conservation Science and Engineering (Springer Nature) and Chemical Engineering Transactions (Italian Association of Chemical Engineering). He is the winner of the Innovator of the Year Award 2009 of IChemE, the Young Engineer Award 2010 of IEM, the Outstanding Young Malaysian Award 2012 of Junior Chamber International (JCI), the SCEJ (Society of Chemical Engineers, Japan) Award for Outstanding Asian Researcher and Engineer 2013, the Vice-Chancellor's Achievement Award 2014 (University of Nottingham), and the Top Research Scientist Malaysia 2016 (Academy of Science Malaysia).

processes

MDPI

Editorial

Special Issue on "Process Design, Integration, and Intensification"

Dominic C. Y. Foo [1],* and Mahmoud El-Halwagi [2],*

[1] Department of Chemical and Environmental Engineering, University of Nottingham, 43500 Semenyih, Selangor, Malaysia
[2] Chemical Engineering Department, Texas A&M University, College Station, TX 77843, USA
* Correspondence: Dominic.Foo@nottingham.edu.my (D.C.Y.F.); el-halwagi@tamu.edu (M.E.-H.)

Received: 28 March 2019; Accepted: 28 March 2019; Published: 3 April 2019

With the growing emphasis on enhancing the sustainability and efficiency of industrial plants, process integration and intensification are gaining additional interest throughout the chemical engineering community. Some of the hallmarks of process integration and intensification include a holistic perspective in design, and the enhancement of material and energy intensity. The techniques can apply to individual unit operations, multiple units, a whole industrial facility, or even a cluster of industrial plants.

This Special Issue on "Process Design, Integration, and Intensification" aims to cover recent advances in the development and application of process integration and intensification. Two works related to process design and integration were reported for simultaneous optimisation of water and energy usage in hydraulic fracturing [1], as well as the design of a palm oil milling process [2]. Besides, two works reported process intensification involving desalination unit [3] and reactive distillation [4].

Brief Synopsis of Papers in the Special Issue

In the work of Oke et al. [1], a mathematical model was proposed for simultaneous optimisation of water and energy usage in hydraulic fracturing. The recycling/reuse of fracturing water is achieved through the purification of flowback wastewater using thermally driven membrane distillation (MD). The study also examines the feasibility of utilising the co-produced gas as a potential source of energy for MD. The proposed framework aids in understanding the potential impact of using scheduling and optimisation techniques to address flowback wastewater management.

Foong et al. [2] on the other hand, proposed a hybrid approach to solve a palm oil milling process. The hybrid approach consists of mathematical programming and graphical techniques. The former is used to optimise a palm oil milling process to achieve maximum economic performance. On the other hand, a graphical approach known as feasible operating range analysis (FORA) is used to study the utilisation and flexibility of the developed design.

In the work reported by Alghamdi et al. [3], an integrated study of modeling, optimization, and experimental work was undertaken for a solar-driven humidification and dehumidification desalination system in Saudi Arabia. Design, construction, and operation are performed, and the system is analyzed at different circulating oil and air flow rates to obtain the optimum operating conditions.

The work of Yamaki et al. [4] reported process intensification involving a reactive distillation column. The authors clarified the factors that are responsible for reaction conversion improvement for reactive distillation column used in the synthesis of tert-amyl methyl ether (TAME). The study also analysed the effect of the intermediate reboiler duty on the reaction performance. The results revealed that the liquid and vapor flow rates influenced the reaction and separation performances, respectively.

Another work that investigated the improvement on the chemical reaction was reported by Yang et al. [5], who proposed an optimisation methodology using Computational Fluid Dynamics (CFD) based compartmental modelling to improve mixing and reaction selectivity. Results demonstrate that reaction selectivity can be improved by controlling rates and feed locations of the reactor. The proposed approach was demonstrated with Bourne competitive reaction network.

The adsorptive properties of poly(1-methylpyrrol-2-ylsquaraine) (PMPS) particles were investigated by Ifelebuegu et al. [6]. The PMPS particles were synthesised by condensing squaric acid with 1-methylpyrrole in butanol, and serves as an alternative adsorbent for treating endocrine-disrupting chemicals in water. The results demonstrated that PMPS particles are effective in the removal of endocrine disrupting chemicals (EDCs) in water, though the removal process was complex and involves multiple rate-limiting steps and physicochemical interactions between the EDCs and the particles.

Abdullah et al. [7] proposed some techniques for improving the reliability of predictive functional control (PFC), when the latter is applied to systems with challenging dynamics. Instead of eliminating or cancelling the undesirable poles, this paper proposes to shape the undesirable poles in order to further enhance the tuning, feasibility, and stability properties of the PFC. The proposed modification is analysed and evaluated on several numerical examples and also a hardware application.

In the perspective paper by Uhlenbrock et al. [8], business models and the regulatory framework regarding the extraction of traditional herbal medicines as complex extracts are outlined. Accordingly, modern approaches to innovative process design methods are necessary. Besides, the benefit of standardised laboratory equipment combined with physico-chemical predictive process modeling, and innovative modular, flexible manufacturing technologies—which are fully automated by advanced process control methods, are described.

Prof. Dr. Mahmoud El-Halwagi
Prof. Dr. Dominic C. Y. Foo
Guest Editors

References

1. Oke, D.; Majozi, T.; Mukherjee, R.; Sengupta, D.; El-Halwagi, M. Simultaneous Energy and Water Optimisation in Shale Exploration. *Processes* **2018**, *6*, 86. [CrossRef]
2. Foong, S.; Andiappan, V.; Tan, R.; Foo, D.; Ng, D. Hybrid Approach for Optimisation and Analysis of Palm Oil Mill. *Processes* **2019**, *7*, 100. [CrossRef]
3. Alghamdi, M.; Abdel-Hady, F.; Mazher, A.; Alzahrani, A. Integration of Process Modeling, Design, and Optimization with an Experimental Study of a Solar-Driven Humidification and Dehumidification Desalination System. *Processes* **2018**, *6*, 163. [CrossRef]
4. Yamaki, T.; Matsuda, K.; Na-Ranong, D.; Matsumoto, H. Intensification of Reactive Distillation for TAME Synthesis Based on the Analysis of Multiple Steady-State Conditions. *Processes* **2018**, *6*, 241. [CrossRef]
5. Yang, S.; Kiang, S.; Farzan, P.; Ierapetritou, M. Optimization of Reaction Selectivity Using CFD-Based Compartmental Modeling and Surrogate-Based Optimization. *Processes* **2019**, *7*, 9. [CrossRef]
6. Ifelebuegu, A.; Salauh, H.; Zhang, Y.; Lynch, D. Adsorptive Properties of Poly(1-methylpyrrol-2-ylsquaraine) Particles for the Removal of Endocrine-Disrupting Chemicals from Aqueous Solutions: Batch and Fixed-Bed Column Studies. *Processes* **2018**, *6*, 155. [CrossRef]
7. Abdullah, M.; Rossiter, J. Input Shaping Predictive Functional Control for Different Types of Challenging Dynamics Processes. *Processes* **2018**, *6*, 118. [CrossRef]
8. Uhlenbrock, L.; Sixt, M.; Tegtmeier, M.; Schulz, H.; Hagels, H.; Ditz, R.; Strube, J. Natural Products Extraction of the Future—Sustainable Manufacturing Solutions for Societal Needs. *Processes* **2018**, *6*, 177. [CrossRef]

processes

MDPI

Article

Simultaneous Energy and Water Optimisation in Shale Exploration

Doris Oke [1] , **Thokozani Majozi** [1,*], **Rajib Mukherjee** [2], **Debalina Sengupta** [2] and
Mahmoud M. El-Halwagi [3]

1   School of Chemical and Metallurgical Engineering, University of the Witwatersrand, 1 Jan Smuts Avenue,
     Braamfontein, Johannesburg 2000, South Africa; funmmydoris@gmail.com
2   Gas and Fuels Research Center, Texas A&M Engineering Experiment Station, College Station,
     TX 77843, USA; rmukhe0@gmail.com (R.M.); debalinasengupta@tamu.edu (D.S.)
3   Chemical Engineering Department, Texas A&M University, College Station, TX 77843-3122, USA;
     el-halwagi@tamu.edu
*   Correspondence: thokozani.majozi@wits.ac.za; Tel.: +27-117176517

Received: 5 May 2018; Accepted: 3 July 2018; Published: 6 July 2018

Abstract: This work presents a mathematical model for the simultaneous optimisation of water and energy usage in hydraulic fracturing using a continuous time scheduling formulation. The recycling/reuse of fracturing water is achieved through the purification of flowback wastewater using thermally driven membrane distillation (MD). A detailed design model for this technology is incorporated within the water network superstructure in order to allow for the simultaneous optimisation of water, operation, capital cost, and energy used. The study also examines the feasibility of utilising the co-produced gas that is traditionally flared as a potential source of energy for MD. The application of the model results in a 22.42% reduction in freshwater consumption and 23.24% savings in the total cost of freshwater. The membrane thermal energy consumption is in the order of 244×10^3 kJ/m^3 of water, which is found to be less than the range of thermal consumption values reported for membrane distillation in the literature. Although the obtained results are not generally applicable to all shale gas plays, the proposed framework and supporting models aid in understanding the potential impact of using scheduling and optimisation techniques to address flowback wastewater management.

Keywords: hydraulic fracturing; water; energy; membrane distillation; optimisation

1. Introduction

The "shale revolution" has triggered a dramatic change in oil and natural gas production globally. From 2007 to 2015, the US witnessed an increase in the amount of shale gas produced from 2 to 15 trillion cubic feet per year [1], with estimates of continued growth to support monetisation projects [2]. The process by which shale gas production is carried out, known as hydraulic fracturing, is associated with several environmental challenges, i.e., water usage and wastewater discharge as well as flaring of co-produced gas. Water management decisions within shale gas production can be grouped into two main categories, i.e., the usage of water in the process of hydraulic fracturing and managing the effluent generated from drilling and production. The production of shale gas typically requires 7000 to 18,000 m^3 of water to fracture and drill a typical well [3–5]. A main challenge associated with water usage in hydraulic fracturing is the relatively short time within which the large volume of fracturing fluid is needed [4]. Another issue of contention that has impeded the ongoing progress in shale gas production processes is water contamination. Two categories of wastewater are generated: flowback water and produced water. Flowback water is the wastewater that returns to the surface within the first few weeks after hydraulic fracturing, and is characterised by a high flowrate and volume generated in the range between 10% and 40% of the initial injected fluid [4]. The contaminants found in flowback water include total suspended solids (TSS), metals, organics, and total dissolved solids (TDS), with the TDS value ranging between 20,000 mg/L and 300,000 mg/L depending on the shale formation and how long the water remains underground [3,4]. Produced water, on the other hand, is the wastewater generated in the production stages. It is made up of the formation water and the injected fracturing fluid generally characterised by high salinity. In selecting appropriate options for the effective management of the high volume of the generated flowback water, several factors have to be considered. These include environmental regulation, the amount and types of contaminants in the wastewater, and economics factors. Thus, water consumption in shale gas production is a serious matter, making water resource management an important operational and environmental issue [6]. The increase in the cost of freshwater and the disposal of generated wastewater, limitations in providing fresh water for fracturing, and the concerns about the negative environmental impact of shale gas wastewater have spurred the interest in identifying cost-effective technologies that can reduce the usage of fresh water and the discharge of wastewater in shale gas production [7].

The proper management of water resources requires wastewater treatment for reuse and/or recycling, which can be accomplished by the use of water treatment units, categorised as membrane or non-membrane processes. Flowback water reuse in hydraulic fracturing demands low salinity, as high salinity can lead to formation damage, affect the performance of some friction reducers, and damage the drilling equipment [8]. The choice of the treatment technology depends on the level of purity required, the mobility, and the economics of the process. The membrane-based process for water treatment is energy intensive; therefore, minimising energy is also of great importance. In this study, we considered membrane distillation (MD) as the membrane technology of interest. MD has emerged as a promising technology in wastewater treatment, gaining a high level of interest in industries especially where high purity separation is of great importance. It is capable of treating wastewater from oil and gas effectively [8]. In MD, the feed is pre-heated to a temperature below the boiling point, which ranges between 323 and 363 K in the case of water treatment application. The water vapour then travels through a hydrophobic, microporous membrane. The vapour is condensed on the permeate side using the stored permeate and collected as pure liquid. The driving force in membrane distillation is the chemical potential difference across the membrane, which depends on the difference between the vapour pressure of the feed and the permeate sides. There are various benefits associated with the use of MD in the areas of water recycling and/or reuse as well as desalination, particularly in shale exploration [8,9]. These include:

- Low-level heating and the ability to operate with moderate temperature and pressure; this is a very crucial factor in shale exploration due to the availability of wasted energy from flaring which can be used as an energy source for MD.
- The ability to treat a highly concentrated feed, which is the case with water, generated from hydraulic fracturing.
- Compact size and modular nature: MD is characterised with a small footprint due to the high surface area to volume ratio of the membrane. It can also be easily adjusted to the required capacity by the removal or addition of MD modules, which allow for easy movement from one well pad to another. All of these factors make MD a candidate desalination technology in this study.

Several authors have developed various optimisation strategies for water management in shale gas production. Yang et al. [4] developed a mixed integer linear programming (MILP) model, which later extended to a mixed integer nonlinear programming (MINLP) model [10] for the investment and scheduling of optimal water management in shale gas production using a discrete time formulation. The linear and nonlinear models dealt with short-term and longer-term operations, respectively. Gao and You [11] approached a similar issue by developing a mixed integer linear fractional programming (MILFP) that focuses on the minimisation of freshwater use in hydraulic fracturing per unit of profit but assumed a fixed schedule for the well pad fracturing. Gao and You [12] also developed a stochastic mixed integer linear fractional programming (SMILFP) model for the optimisation of the levelized cost of energy produced from shale gas. Elsayed et al. [8] proposed an optimisation method based on multi-period formulation for the treatment of shale gas flowback water, which takes into account the fluctuation in the contaminant concentration and flowrate using membrane distillation. Bartholomew and Mauter [13] developed a multi-objective MILP model which is formulated to determine the water management approach that reduces both financial, human health, and environment cost associated with shale gas water management. Lira-Barragán et al. [14] developed an optimisation framework to deal with the uncertainties associated with the management of water in shale gas production. However, most of these studies have either adopted the discrete time scheduling formulation for the well pad fracturing or assumed a fixed schedule. A limitation of discretising the time horizon is the explosive binary dimension that could lead to higher computational time and suboptimal solution. Assuming a fixed schedule is a huge drawback, as this has a great effect on the overall profit. In addition, most of the research conducted in this area has represented the wastewater treatment unit as a "black box" which does not give the true cost representation of the project or uses "short cut" regenerator model [15] due to the complexity of the regenerator design.

Flaring is the burning of natural gas that cannot be refined or sold. Flaring is carried out frequently in most industrial plants, especially in managing unusual or irregular occurrences. Flaring in most industries is carried out to decrease hazard in the course of distress in an industrial operation, to get rid of associated gases, or to safely manage process start-up and shutdown [16]. In order to minimise flaring in industries, legislative acts should be implemented so that industries will take necessary precautions. Another way of achieving this is by the recovery and efficient utilisation of flaring streams [17]. In the context of shale exploration, flaring is common in areas where oil and gas are co-produced with no sufficient infrastructure for gathering the gas. Because of this drawback, the producer either choses to build the pipeline or gathering facilities, flare the gas, or find a useful way of utilising the gas on site [18]. Although facts about the rate of flaring after well completion is yet to be published, information from the literature suggests that the time at which gas is mostly flared coincides with the time when a substantial volume of flowback water is recovered. According to Glaizer et al. [18], flaring of gas is mostly done in the first 10 producing days after initial completion or recompletion of a well. For example, 15,041 wells were completed in Texas in 2012, which led to the flaring of 1.36 billion m^3 (48 billion ft^3) of natural gas. The estimated rate of flare based on these figures can be set at 9600 m^3 per well per day, though variation might occur based on a particular well [18]. In general, flaring is found to be a waste of resources globally, resulting in serious environmental problems such as air, thermal, and light pollution [19]. Studies available in the literature for the utilisation of the co-produced gas that is flared after well completion is either focused on onsite atmospheric water harvesting [19] using the captured gas or using it as a source of heat [18] for heat-based regenerators. However, it needs to be mentioned that the work by Glazer et al. [18] was conducted based on analytical framework and not in the context of mathematical optimisation.

This paper focuses on the synthesis and optimisation of an integrated water and membrane network that simultaneously optimises water and energy consumption in hydraulic fracturing using continuous time mathematical formulation for scheduling. The membrane technology considered is membrane distillation (MD). A detailed design of MD is incorporated to determine the optimal operating conditions for efficient energy use. The rest of the paper is structured as follows. Section 2 gives the general problem statement and its assumption. Section 3 provides detailed information about the superstructure for the total network. The model formulation is presented in Section 4, while in Section 5 a case study is examined to demonstrate the model applicability. Finally, the conclusions are given in Section 6.

2. Problem Statement

The problem statement in this work can be stated as follows.

Given the following:

- Number of freshwater sources (interruptible and uninterruptible);
- Set of well pads S to be fractured with a known volume of water required for fracturing and a maximum allowable contaminant concentration in the fracturing fluid;
- Total number of frac stages for each well pad;
- Earliest fracturing date for each well pad;
- Set of wastewater injection wells D;
- Volume of water required per stage;
- Minimum and maximum number of stages that can be fractured per day;
- Time horizon of interest;
- Network of regenerator;
- Gas storage facility;
- Historical stream data for the interruptible source,

Determine the optimal configuration of the total network that gives:

- Optimal fracturing schedule of the well pads;
- Minimum freshwater intake and wastewater generation;
- Optimal operation and design conditions of the regenerator such as the number of membrane modules and the energy consumption;
- Feasibility of using captured flared gas as an energy source for the regenerator.

The assumptions made in the model formulation are as follows:

- The wells in each well pad are aggregated [4];
- Each well pad is connected to exactly one of the impoundment through piping [4];
- The number of fracturing stages that could be fractured per day is kept constant at 4 instead of allowing it to be variable between 2 and 4 stages [4];
- The flowback water from the fractured well pad is assumed to be 25% [10] of the initial water used;
- The capacity of the wastewater tank and fracturing tank on each well pad varies depending on its water requirement;
- The water treatment unit is located onsite and can be moved from one well pad to the other;
- The historical flowrate data for the interruptible water source from each calendar year is treated as a scenario, and each year is treated with equal probability [4].

3. Superstructure Representation

Based on the problem statement, the superstructure in Figure 1 is developed. In the superstructure, two types of freshwater sources are considered (interruptible and uninterruptible sources) [4]. An uninterruptible source is a big water body with guaranteed water availability throughout the year, but the mode of transportation is trucking. The interruptible source is a nearby source that requires piping but with uncertain water availability all year round. These two sources are considered because water management decisions are primarily influenced by transportation costs [4]. In order to complete a typical well pad, roughly 4000–6500 one-way truck trips are needed. Hence, due to the high cost of trucking and other environmental impacts related to drawing water from uninterruptible sources, operators are encouraged to draw water from sources that are close by through piping [4]. The water from any of these sources can be stored in any impoundment t prior to its usage. S represents a set of well pads to be fractured in which the fracking fluid is blended using freshwater from the impoundment and the recycled water from the fracturing tank. The maximum concentration of TDS into the well pads is kept at an upper limit of 50,000 ppm [10,13]. The flowback water generated from the fractured well pad in the first two weeks after fracturing is assumed to be 25% of the initial water used [10]. This flowback water can be sent to regenerator R for treatment or any injection well D for disposal. The flowback water sent to regenerator R is treated before it is sent to the fracturing tank for reuse in the next well pad. The product of a particular well pad after stimulation can be either oil and gas or gas only, depending on the geological formation of the shale play. For a well pad that produces oil and gas, the co-produced gas can be captured and stored in the gas storage facility from where it is supplied to the regenerator R as fuel, which in turn produces the heat energy needed by the regenerator while the oil is sent to the market. In the case of a gas-producing well, part of the gas can be diverted into the gas storage facility for wastewater treatment while the rest can be sent to the market.

The mode of operation of the regenerator is as stated below:

- The transfer of water from the wastewater tank to the regenerator R is conducted provided that there is a well pad to be fractured. Whenever the regenerator starts operation, it operates continuously until the wastewater tank becomes empty.
- The regenerator only operates if there is a well pad to be fractured, otherwise it remains inactive.
- The performance of the regenerator is specified based on a variable removal ratio.

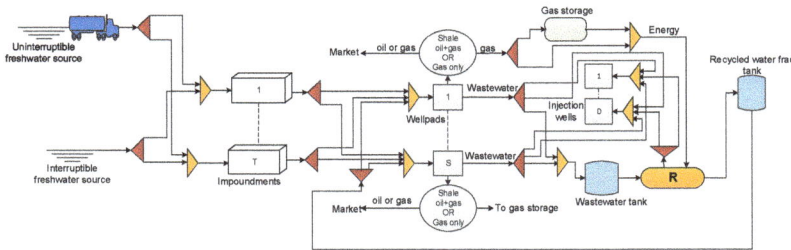

Figure 1. Superstructure representation of the water network.

4. Model Formulation

The mathematical model presented in this section is based on the superstructure given in Figure 1. The problem is formulated as a mixed integer nonlinear programming (MINLP) model, which is divided into two sections developed inside the same structure to simultaneously optimise water and energy. The first section focuses on mass balance and scheduling while the second is based on the detailed membrane distillation model. The scheduling framework adopted here is based on the state task network (STN) and unequal discretisation of the time horizon, which involves time point n occurring at an unknown time. A time point is a precise moment within a given horizon when an event occurs (e.g., start of task, end of task, transfer of materials, etc.). It is generally used to track inventory levels and model the occurrence of tasks in batch and semi-batch processes. Among the important decision variables are the 0–1 variables which indicate if a well pad is fractured or if water is transferred to storage and if regeneration takes place. The following three sets of binary variables are used:

$w_{s,n}$ is assigned a value of 1 if well pad s is stimulated at time point n.

$wv_{s,n}$ is assigned a value of 1 if the transfer of water takes place from well pad s to storage at time point n.

wr_n is assigned a value of 1 if the transfer of water from storage to the regenerator takes place at time point n.

In order to explain the model, the constraints characterising the optimisation formulation are described.

4.1. Mass Balance Constraint

It is important to state the mass balances around each well pad, the impoundment, the wastewater storage tank, the fracturing tank, the injection well, and the regenerator.

4.1.1. Mass Balance around Well Pad *s*

The mass balance around a well pad is conducted in accordance with Figure 2. The total volume of water required to fracture well pad *s* at time point *n*, $f_{s,n}$, is given by Equation (1), where WR_s is the amount of water required to fracture well pad *s* and $w_{s,n}$ is the binary variable that indicates if well pad *s* is fractured at time point *n*. This water requirement is supplied with freshwater from the impoundment $f_{s,n}^{fw}$ and/or reused water from the fracturing tank $f_{s,n}^{ww}$, which is obtained by Equation (2). Equation (3) specifies that only freshwater is to be used at the first time point.

$$f_{s,n} = WR_s w_{s,n} \quad \forall s \in S, \, n \in N \tag{1}$$

$$f_{s,n} = f_{s,n}^{fw} + f_{s,n}^{ww} \quad \forall s \in S, \, n \in N, n \geq 2 \tag{2}$$

$$f_{s,n} = f_{s,n}^{fw} \quad \forall s \in S, \, n \in N, n = 1 \tag{3}$$

The flowback water generated in the first two weeks after fracturing $f_{s,n}^{fbw}$ is assumed to be 25% of the initial water used and is given by Equation (4). Equation (5) gives the TDS concentration $c_{s,n}^{fbw}$ in the wastewater where CS_s is the flowback water concentration of well pad *s*. The value used is between the average value in the first two weeks after fracturing and the highest value that can be found in typical flowback water, as reported in literature. Equation (6) states that the flowback water after well pad fracturing could be discarded as effluent or sent to the wastewater storage tank where $f_{s,n}^{st}$ is the volume of wasewater sent to storage and $f_{s,n}^{dis}$ is the volume of wastewater sent to disposal from well pad *s* at time point *n*.

$$f_{s,n}^{fbw} = 0.25 f_{s,n} \quad \forall s \in S, \, n \in N \tag{4}$$

$$c_{s,n}^{fbw} = CS_s w_{s,n} \quad \forall s \in S, \, n \in N \tag{5}$$

$$f_{s,n}^{fbw} = f_{s,n}^{st} + f_{s,n}^{dis} \quad \forall s \in S, \, n \in N \tag{6}$$

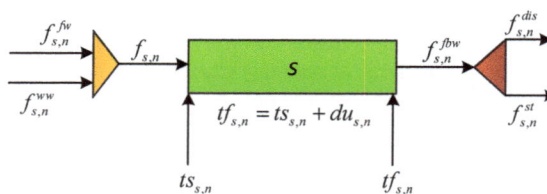

Figure 2. Mass balance representation around a well pad.

The mass balance around the impoundment is conducted in accordance with Figure 3, as given in Equations (7) and (8). Equation (7) describes the total water use $it_{t,n}^{fw}$ from impoundment *t* at time point *n* given the piping connection $TP_{s,t}$ between impoundment *t* and well pad *s*. The volume $vit_{t,n,y}$ of impoundment *t* at time point *n* for a given scenario year *y* is described by Equation (8). The equation states that the volume of freshwater stored in the impoundment consists of the volume stored at the previous time point and the difference between the amount of water entering the impoundment through trucking and piping and the total water leaving the impoundment to well pads. $f_{t,n,y}^{pump}$ is a continuous variable which specifies the amount of water supplied through piping from an interruptible source to the corresponding impoundment at time point n and $f_{t,n,y}^{truck}$ is the amount of water supplied through trucking.

$$i_{t,n}^{fw} = \sum_{s \in TP_{s,t}} f_{s,n}^{fw} \quad \forall t \in T, n \in N \tag{7}$$

$$vi_{t,n,y} = vi_{t,n-1,y} + f_{t,n,y}^{pump} - i_{t,n}^{fw} + f_{t,n,y}^{truck} \quad \forall t \in T, n \in N, y \in Y \tag{8}$$

Equation (9) states that the total volume of water disposed fd_n at time point n is the sum of the flowback water sent to disposal $f_{s,n}^{dis}$ from well pad s and the concentrate from the regenerator f_n^{con}. This total amount of water can be disposed into any injection well d, as given in Equation (10), while Equation (11) states that the throughput into each injection well should not exceed the maximum it can take. ff_n^{dis} is a continous variable indicating the throughput of an injection well d at time point n, and DI^{max} is the parameter indicating the maximum capacity of the injection well.

$$fd_n = \sum_s f_{s,n}^{dis} + f_n^{con} \quad \forall n \in N \tag{9}$$

$$fd_n = \sum_d ff_{d,n}^{dis} \quad \forall n \in N \tag{10}$$

$$fd_n \leq DI^{max} \quad \forall n \in N \tag{11}$$

Equation (12) gives the expected production $p_{s,n}$ from well pad s at time point n, where p_s is a parameter indicating the gas production of well pad s.

$$p_{s,n} = P_s w_{s,n} \quad \forall s \in S, n \in N \tag{12}$$

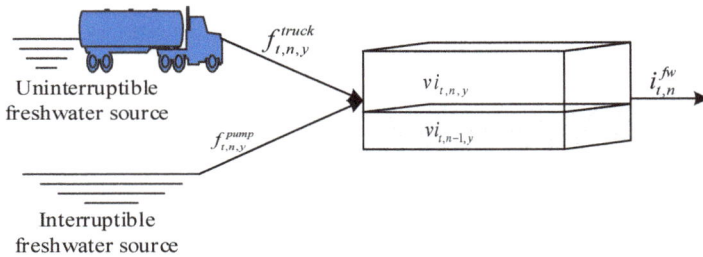

Figure 3. Mass balance representation around the impoundment.

4.1.2. Mass Balance around the Wastewater Storage Tank and the Fracturing Tank

The mass and contaminant balances around the wastewater storage tank and the fracturing tank are conducted in accordance with Figure 4. Part of the assumption made in this study is that all of the flowback water sent to storage from well pad s fractured at a previous time point $f_{s,n-1}^{st}$ is the quantity that is treated by the regenerator f_n^{reg} at time point n, as stated in Equation (13). This indicates that the volume of the wastewater tank on each well pad becomes zero at the end of each time point. The concentration of water sent to the treatment unit is given by Equation (14), where $c_n^{st,ww}$ is the contaminant concentration in the treatment unit at time point n.

$$\sum_s f_{s,n-1}^{st} = f_n^{reg} \quad \forall n \in N, n \geq 2 \tag{13}$$

$$\sum_s f_{s,n-1}^{st} c_{s,n-1}^{fbw} = f_n^{reg} c_n^{st,ww} \quad \forall n \in N, n \geq 2 \tag{14}$$

The capacity of the treatment wastewater tank v_n^{ww} at time point n is bounded by the volume of flowback water f_n^{reg} to be treated at time point n, as given in Equation (15). Equations (16)–(18) ensure that the maximum capacity of the tank is not exceeded, where $V^{\max}$ and $V^{\min}$ are parameters that indicate the maximum and minimum capacity of the wastewater storage tank, respectively. Equation (19) ensures that no water is stored in the storage tank at the end of the time horizon.

$$v_n^{ww} = f_n^{reg} \quad \forall n \in N \tag{15}$$

$$v_n^{ww} \leq V^{\max} \quad \forall n \in N \tag{16}$$

$$f_{s,n-1}^{st} \geq V^{\min} wv_{s,n-1} \quad \forall s \in S, n \in N, n \geq 2 \tag{17}$$

$$f_{s,n-1}^{st} \leq V^{\max} wv_{s,n-1} \quad \forall s \in S, n \in N, n \geq 2 \tag{18}$$

$$v_n^{ww} = 0 \quad \forall n = /N/ \tag{19}$$

The capacity of the fracturing tank $v_{s,n}^{ft}$ on well pad s depends on the volume of wastewater required $f_{s,n}^{ww}$ at the well pad, as defined in Equation (20).

$$v_{s,n}^{ft} \geq f_{s,n}^{ww} \quad \forall s \in S, n \in N \tag{20}$$

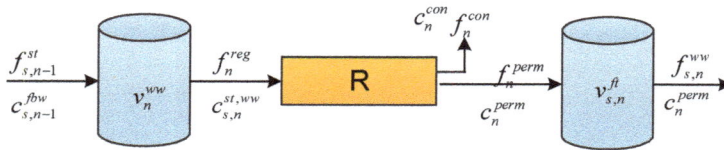

Figure 4. Mass balance representation around the storage tank, regenerator, and fracturing tank.

4.1.3. Mass Balance around the Regenerator

Equation (21) states that the total volume of water into the regenerator at time point n f_n^{reg} is the sum of the amount collected as permeate f_n^{perm} and the amount sent to disposal as concentrate f_n^{con}. The contaminant balance around the regenerator is given in Equation (22), where c_n^{perm} represents the outlet concentration of contaminants from the regenerator and c_n^{con} is the contaminant concentration removed from the water by the regenerator at time point n. Equation (23) states that the inlet contaminant concentration into the regenerator should not exceed the maximum it can take, where $C^{\max}$ is the maximum inlet concentration into the regenerator. The performance of the regenerator is a function of the removal ratio (RR) of contaminants, as stated in Equation (24). The quantity of water to be collected as permeate and concentrate depends on the recovery ratio (LR), as stated in Equations (25) and (26), respectively.

$$f_n^{reg} = f_n^{perm} + f_n^{con} \quad \forall n \in N \tag{21}$$

$$f_n^{reg} c_n^{st,ww} = f_n^{perm} c_n^{perm} + f_n^{con} c_n^{conc} \quad \forall n \in N \tag{22}$$

$$c_n^{st,ww} \leq C^{\max} \quad \forall n \in N \tag{23}$$

$$c_n^{perm} = c_n^{st,ww}(1 - RR) \quad \forall n \in N \tag{24}$$

$$LR f_n^{reg} = f_n^{perm} \quad \forall n \in N \tag{25}$$

$$f_n^{con} = (1 - LR) f_n^{reg} \quad \forall n \in N \tag{26}$$

The amount of wastewater reused at any time point is supplied through the permeate stream from the regenerator, as given in Equation (27). Equation (28) ensures that the maximum allowable concentration in the well pad is not exceeded, where CS^{max} is the maximum inlet contaminant concentration in well pad s.

$$f_n^{perm} = \sum_s f_{s,n}^{ww} \quad n \in N \tag{27}$$

$$f_n^{perm} c_n^{perm} \leq CS^{max} \sum_s f_{s,n} \quad \forall n \in N \tag{28}$$

4.2. Scheduling Model

The scheduling part of the model captures the time dimension related to the process. These are categorised into three parts, namely:

- well pad scheduling,
- wastewater storage tank scheduling, and
- regenerator scheduling.

4.2.1. Well Pad Scheduling

Equation (29) is the allocation constraint that specifies that each well pad s has to be fractured exactly once at a given time point n in the time horizon.

$$\sum_n w_{s,n} = 1 \quad \forall s \in S \tag{29}$$

Equation (30) states that no task can start at the end of the time horizon.

$$w_{s,n} = 0 \quad \forall s \in S, n \in N, n = /N/ \tag{30}$$

The duration of each well pad $du_{s,n}$ is calculated by Equation (31), where TR_s is the time required to fracture well pad s. Equations (32) and (33) give the finish time of each well pad $tf_{s,n}$ expressed with big-M constraints, which are only active if well pad s is stimulated at time point n, where $ts_{s,n}$ is the start time of fracturing well pad s at time point n.

$$du_{s,n} = TR_s w_{s,n} \quad \forall s \in S, n \in N \tag{31}$$

$$tf_{s,n} \leq ts_{s,n} + du_{s,n} + H(1 - w_{s,n}) \quad \forall s \in S, n \in N \tag{32}$$

$$tf_{s,n} \geq ts_{s,n} + du_{s,n} - H(1 - w_{s,n}) \quad \forall s \in S, n \in N \tag{33}$$

Equation (34) states that the time at which the fracturing of well pad s begins, $ts_{s,n}$, is equal to the time at which time point n occurs tt_n, i.e., the start time of each well pad must coincide with a time point.

$$ts_{s,n} = tt_n \forall s \in S, n \in N \tag{34}$$

Equation (35) gives the sequence-dependent change over time between well pad s and s'. It states that the start time of well pad s' at time point n must be equal to or greater than the finish time of well pad s at a previous time point plus the crew transition time $CT_s^{s'}$ between well pad s and s'. Equation (36) states that the time at which time point n occurs must correspond with the availability time AT_s of well pad s.

4.2.2. Storage Tank Scheduling

$$ts_{s',n} \geq tf_{s,n-1} + CT_s^{s'} w_{s',n} \quad \forall s \in S, s' \in S, s' \neq s, n \in N, n \geq 2 \tag{35}$$

$$tt_n \geq \sum_s (AT_s w_{s,n} - H(1 - w_{s,n})) \quad n \in N \tag{36}$$

Water usage in hydraulic fracturing and the water sent to the storage tank for treatment are linked by Equation (37). This equation states that water can only be transferred from well pad s to the wastewater tank for treatment if well pad fracturing takes place at that time point. Equations (38) and (39) ensure that the transfer time of water from a well pad into storage $tv_{s,n}$ corresponds with the time when the fracturing task ends $tf_{s,n}$.

$$wv_{s,n} \leq w_{s,n} \quad \forall s \in S, n \in N \tag{37}$$

$$tv_{s,n} \geq tf_{s,n} - H(2 - wv_{s,n} - w_{s,n}) \quad \forall s \in S, n \in N \tag{38}$$

$$tv_{s,n} \leq tf_{s,n} + H(2 - wv_{s,n} - w_{s,n}) \quad \forall s \in S, n \in N \tag{39}$$

4.2.3. Regenerator Scheduling

Equation (40) relates the regeneration and fracturing task starting at time point n. It states that water regeneration can only take place at time point n if there is a well pad to be fractured at that time point. Equations (41) and (42) ensure that the time at which regeneration starts tr_n coincides with the time at which the fracturing task starts, at time point n. This is because all tasks starting at point n must start at the same time, although their finish times do not have to coincide. Equation (43) gives the duration of regeneration, where ttr_n is the finish time of regeneration at time point n, f_n^{reg} is the total volume of water in the regenerator at time point n, and ff^{MD} is the feed flowrate into the regenerator.

$$wr_n \geq w_{s,n} \quad \forall s \in S, n \in N \tag{40}$$

$$tr_n \geq ts_{s,n} - H(2 - wr_n - w_{s,n}) \quad \forall s \in S, n \in N \tag{41}$$

$$tr_n \leq ts_{s,n} + H(2 - wr_n - w_{s,n}) \quad \forall s \in S, n \in N \tag{42}$$

$$ttr_n = tr_n + \left(\frac{f_n^{reg}}{ff^{MD}} \right) wr_n \quad \forall n \in N \tag{43}$$

4.2.4. Tightening Constraint

Tightening formulations play an important role in finding good solutions for the original problem. Not adding a tightening constraint can lead to weak relaxation. Equation (44) imposes the requirement that the sum of the fracturing durations of all well pads $du_{s,n}$ should be less than or equal to the time horizon H, while Equation (45) restricts the sum of the fracturing time of all well pads starting after tt_n to be smaller than the remaining time, where tt_n is the time at which time point n occurs.

$$\sum_s \sum_n du_{s,n} \leq H \tag{44}$$

$$\sum_s \sum_{n' \geq n} du_{s,n'} \leq H - tt_n \quad \forall n \in N \tag{45}$$

4.3. Membrane Distillation (MD) Model

The detailed design model for the membrane distillation unit, which is based on the work of Elsayed et al. [9], is presented in this section. Various configurations of MD have been reported in the literature [9,20] with variation based on mode of vapour collection on the permeate side and the method of the driving force enhancement across the membrane. The focus of this study is on direct contact membrane distillation (DCMD), which is found to be the most commonly used configuration. Some of the merits associated with DCMD are ease of construction, operation, maintenance, and stability in operation [9]. Figure 5 illustrates a schematic representation of a typical direct contact membrane distillation unit. The flowback water is pre-heated to effect evaporation and the degree of pre-heating becomes an optimisation variable. The vapour passes through the membrane and condensation occurs on the permeate side using stored permeate, which is at relatively low temperature than the feed. Consideration must be given to both heat and mass transfer from the feed side to the permeate side of the membrane. Mass and heat transfer takes place across three sections [9,20]: mass transfer takes place in the boundary layer of the membrane on the feed side, across the membrane, and on the permeate side boundary layer. Heat transfer, on the other hand, takes place from the bulk of the feed to the interface of the membrane through a boundary layer via convection, across the membrane via conduction and latent heat associated with the vaporised flux, and through the boundary layer from the interface of the membrane to the bulk of the permeate via convection. In order to describe mass transfer through the membrane, a model such as Knudsen diffusion, molecular diffusion, or the incorporation of both have been established to yield quality results [20].

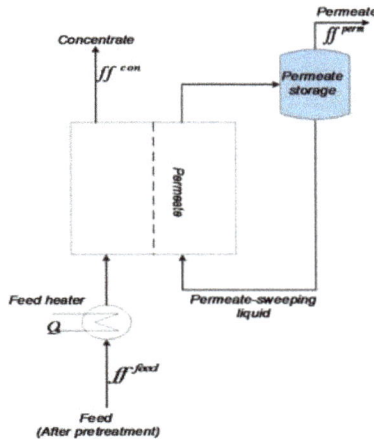

Figure 5. Schematic representation of a typical direct contact membrane distillation.

The membrane distillation considered is a polyvinylidene fluoride flat sheet membrane used in DCMD. The details of this are given in Yun et al. [20].

The following assumptions are made for the constraints in the plant using a set of mathematical equations describing its operation:

Flowback water contains organics, oils, and total dissolved solids (TDS), mainly in the form of salts and other contaminants [21]. It is assumed that the flowback water is pre-treated to remove oils, organics, and other necessary contaminants. Membrane distillation is used to remove TDS, as this is the main contaminant of interest for water reuse/recycling in hydraulic fracturing.

The separation efficiency of the MD modules depends on temperature. This is because the permeate flux is also temperature-dependent.

The driving force for the water flux across the membrane, Jw, is the difference in pressure of the water vapour and is defined in Equation (46):

$$Jw = Bw\left(p_{wf}^{vap}\gamma_{wf}x_{wf} - p_{wp}^{vap}\right) \tag{46}$$

where Bw is the membrane permeability, p_{wf}^{vap} is the water vapour pressure of the feed, p_{wp}^{vap} is the water vapour pressure of the permeate, γ_{wf} is the activity coefficient of water in the feed, and x_{wf} is the mole fraction of water in the feed.

The Antoine equation [21] is used to estimate the water vapor pressure of the feed and the permeate which depends on the temperature as given in Equations (47) and (48), where T_{mf} and T_{mp} are the temperature of the feed and the permeate on the membrane, respectively.

$$p_{wf}^{vap} = \exp\left(23.1964 - \frac{3816.44}{T_{mf} - 46.13}\right) \tag{47}$$

$$p_{wp}^{vap} = \exp\left(23.1964 - \frac{3816.44}{T_{mp} - 46.13}\right) \tag{48}$$

The activity coefficient is dependent on the concentration and on the assumption of NaCl as the primary solute. Equation (49) [22] can be used to estimate the activity coefficient, where x_{NaCl} is the molar concentration of NaCl in the feed.

$$\gamma_{wf} = 1 - 0.5x_{NaCl} - 10x_{NaCl}^2 \tag{49}$$

Sodium chloride is chosen as the basis of calculation because it is reported to be the dominant species with regards to the concentration in the flowback/produced water [23–25]. It makes up over 50% of the total dissolved solids.

The permeability of the membrane Bw depends on the membrane temperature T_m, which differs based on the kind of diffusion. The permeability of membranes in which molecular diffusion occurs is calculated through Equation (50) as proposed by Elsayed et al. [9], where B_{wb} is the temperature-independent base value of membrane permeability.

$$Bw = B_{wb}T_m^{1.334} \tag{50}$$

The membrane temperature is the mean value of the bulk temperature of the feed, T_{bf}, and of the permeate, T_{bp} [26]. Therefore, the average temperature in the MD module can be determined by the expression given in Equation (51).

$$T_m = \frac{T_{bf} + T_{bp}}{2} \tag{51}$$

Mass and salt balance around the MD unit is conducted in accordance with Figure 3, as given in Equations (52)–(54):

$$ff^{feed} = ff^{MD}\rho_{water} \tag{52}$$

$$ff^{feed} = ff^{perm} + ff^{con} \tag{53}$$

$$ff^{feed}cf^{feed} = ff^{perm}cp^{perm} + ff^{con}cr^{con} \tag{54}$$

where ff^{feed} is the total flowrate into MD, ff^{perm} and ff^{con} are the permeate and concentrate flowrate, and ρ_{water} is the density of the water. The amount of water collected as permeate highly depends on the energy Q supplied to the unit. Therefore, the heat required by the feed is given in Equation (55):

$$Q = ff^{feed}C_P\left(T_{bf} - T_{sf}\right) \tag{55}$$

where C_p, T_{bf}, T_{sf}, are the specific heat capacity of the feed stream, temperature of the feed in the bulk, and temperature of the feed water into MD, respectively. Only a portion of the heat supplied to the unit is used to vaporise the permeate. This portion is the efficiency factor η. Thus, Equation (56) gives the heat balance for the MD unit [9], where ΔH_{vw} is the latent heat of vaporisation for water.

$$\eta Q = ff^{perm}\Delta H_{vw} \tag{56}$$

The thermal efficiency of MD, η, can be measured using experimental data or a semi-empirical formula [9], as indicated in Equation (57). In this equation, k_m is the membrane thermal conductivity and δ is the membrane thickness.

$$\eta = 1 - \frac{1.5\frac{k_m}{\delta}\left(T_{mf} - T_{mp}\right)}{Jw\Delta H_{vw} + \frac{k_m}{\delta}\left(T_{mf} - T_{mp}\right)} \tag{57}$$

The thermal conductivity of a particular membrane can be determined using Equation (58), which is correlated based on the data of Khayet and Matsuura [27].

$$k_m = 1.7 \times 10^{-7}T_m - 4.0 \times 10^{-5} \tag{58}$$

Equation (59), as proposed by Elsayed et al. [9], can be used to determine water vaporisation in the feed side of the membrane.

$$\Delta H_{vw} = 3190 - 2.5009T_{mf} \tag{59}$$

The transfer of heat through the boundary layers on the two sides of the membrane results in a temperature gradient between the bulk solutions and the surface of the membrane known as temperature polarisation, θ. This occurrence may lead to a significant reduction in the driving force; therefore, it is necessary to consider the temperature gradient across the membrane. Based on this, the temperature polarisation coefficient [28] may be used to calculate the membrane temperature profile as given in Equation (60).

$$\theta = \frac{T_{mf} - T_{mp}}{T_{bf} - T_{bp}} \tag{60}$$

In order to estimate the temperature polarisation coefficient of a particular membrane, experimental data or correlations may be used [27,29]. A linear behaviour as a function of the temperature, as provided by Khayet and Matsuura [27], is given in Equation (61).

$$\theta = 1.362 - 0.0026T_{bf} \tag{61}$$

In accordance with the experimental observation, two other simple assumptions are suggested [9,26]. The first assumption is that for MD with laminar flows of the feed and the sweeping liquid, the absolute value of the temperature difference between the bulk and the membrane on each side of the membrane is nearly the same, as given in Equation (62).

$$T_{bf} - T_{mf} \approx T_{mp} - T_{bp} \tag{62}$$

The second assumption is that the membrane temperature is the mean value of the bulk temperature of the feed and permeate, as specified in Equation (51) above. The liquid recovery, LR, is the fraction of the feed in the regenerator that is recovered as permeate. The fraction of water recovery by the MD unit is given by Equation (63).

$$LR = \frac{ff^{perm}}{ff^{feed}} \tag{63}$$

The removal ratio, RR, is the mass load of contaminants in the concentrate stream of the regenerator as a fraction of the feed. It is assumed to be a variable in this work and is defined as given in Equation (64).

$$RR = \frac{ff^{con}cr^{con}}{ff^{feed}cf^{feed}} \tag{64}$$

In order to determine the area of the membrane required, A_m, the permeate flow rate is divided by the water flux as given in Equation (65).

$$A_m = \frac{ff^{perm}}{Jw} \tag{65}$$

The regeneration network takes into account the capital and the operational cost involved in the operation of the unit. These are incorporated in the overall objective function in order for the energy consumed as well as the cost associated with regeneration to be optimised together with water utilisation. The annual fixed cost of the MD network, AFC, as proposed by Elsayed et al. [9], is given by Equation (66).

$$AFC = 58.5A_m + 1115ff^{feed} \tag{66}$$

The annual operating cost excluding heating, AOC, as proposed by Elsayed et al. [9], is given by Equation (67), where u is the ratio of recycled reject to raw feed.

$$AOC = (1411 + 43(1 - LR) + 1613(1 + u))ff^{feed} \tag{67}$$

The annual heating cost, AHC, is given by Equation (68), where AOT is the annual operating time, Q is the heat requred by the feed into MD, and OC^{ht} is a parameter indicating the cost of heating.

$$AHC = AOT\left(QOC^{ht}\right) \tag{68}$$

4.4. Additional Constraints

The thermal energy consumption per unit of water treated, E^{con}, is given by Equation (69). Equation (70) gives the total energy required for treatment at any time point, E_n^{total}. The volume of natural gas needed per time point, V_n^{nat}, is given in Equation (71), where ∂_{ED} is the energy density.

$$E^{cons} = \frac{Q}{ff^{feed}} \tag{69}$$

$$E_n^{total} = f_n^{reg} E^{cons} \forall n \in N \tag{70}$$

$$V_n^{nat} = \frac{E_n^{total}}{38300\partial_{ED}} \forall n \in N \tag{71}$$

4.5. Objective Function

The objective is to maximise profit, which comprises of the following terms: (1) revenue from gas production, (2) freshwater transportation cost, (3) wastewater treatment cost, (4) disposal cost, (5) wastewater storage cost, and (6) pumping cost to treatment facility, as given in Equation (72).

$$\max profit = SP^{gas} \sum_s \sum_n p_{s,n} \\ - \begin{bmatrix} \left(OC^{truck,fw} \sum_t \sum_n \sum_y \frac{f_{t,n,y}^{truck}}{NY} + OC^{pump,fw} \sum_t \sum_n \sum_y \frac{f_{t,n,y}^{pump}}{NY} \right) \\ + \left(\left(58.5A_m + 1115ff^{feed} \right) + (1411 + 43(1 - LR) + 1613(1 + u))ff^{feed} + AOT\left(QOC^{ht}\right) \right) \\ + \left(OC^{dis} \sum_s \sum_n f_{s,n}^{dis} \right) \\ + \left(OC^{st,ww} \left(\sum_s \sum_n v_{s,n}^{ft} + \sum_n v_n^{ww} \right) \right) \\ + OC^{pump,ww} DS_s \left(\sum_s \sum_n f_{s,n}^{st} + \sum_n f_n^{perm} \right) \end{bmatrix} \tag{72}$$

Equations (1)–(71) constitute the full set of constraints for the optimisation program. In the aforementioned formulation, the following is the list of the decision variables for optimisation:

A_m: Total area of membranes (m^2), defined by Equation (65).

$f_{t,n,y}^{pump}$: Water pumped from an interruptible source at time point n in scenario year y (m^3), defined by Equation (8).

$f_{t,n,y}^{truck}$: Water trucked from an uninterruptible source at time point n in scenario year y (m^3), defined by Equation (8).

$f_{s,n}^{fw}$: Freshwater required to fracture well pad s at time point n (m^3), defined by Equation (2).

$f_{s,n}^{ww}$: Wastewater required to fracture well pad s at time point n (m^3), defined by Equation (2).

f_n^{reg}: Total flowback water to be treated at time point n (m^3), defined by Equations (15).

ff^{MD}: Total flowrate into MD (m^3/day), defined by Equation (43).

$i_{t,n}^{fw}$: Total freshwater required from impoundment t for fracturing at time point n (m^3), defined by Equation (7).

Jw: Water flux across the membrane (kg/(m^2·s)), defined by Equation (46).

p_{wf}^{vap}: Water vapour pressure of the feed in MD (pa), defined by Equation (47).

p_{wp}^{vap}: Water vapour pressure of the permeate in MD (pa), defined by Equation (48).

Q: Heat required by the feed into MD (kJ/day), defined by Equation (55).

RR: Regenerator removal ratio, defined by Equation (64).

T_{mf}: Temperature of the feed on the membrane (K), defined by Equation (60).

T_{mp}: Temperature of the permeate on the membrane (K), defined by Equation (60).

T_m: Membrane average temperature (K), defined by Equation (51).

T_{bf}: Temperature of the feed in the bulk (K), defined by Equation (55).

T_{bp}: Temperature of permeate in the bulk (K), defined by Equation (51).

$vi_{t,n,y}$: Volume of impoundment t at time point n in scenario year y (m^3), defined by Equation (8).

γ_{wf}: Activity coefficient of water in the feed for membrane distillation, defined by Equation (49).

5. Case Study

In order to demonstrate the applicability of the proposed model, an example taken from Yang et al. [4] is considered. This case study represents the typical Marcellus Shale play. The example considered 14 well pads, a time horizon of 540 days, one uninterruptible freshwater source, and two interruptible sources connected to impoundments, as illustrated in Table 1. Thirty years of historical data were provided for the two interruptible sources. The selected membrane distillation is a polyvinylidene fluoride used in direct-contact membrane distillation. The details of this membrane module are given in Yun et al. [20] and Elsayed et al. [9]. The permeability of the membrane is a function of the membrane temperature, which varies depending on the type of diffusion. This is calculated based on molecular diffusion through Equation (50). In order to ensure a complete analysis of the model, three different scenarios are considered. Scenario 1 is the base case which is the water integration without regeneration. Scenario 2 is the case where black box model is used; i.e., water minimisation only and a linear cost function is used to estimate the cost associated with regeneration. Scenario 3 considers water integration involving a detailed regenerator where water and energy are optimised simultaneously.

Table 1. Well pad data [4].

Well Pads	S1	S2	S3	S4	S5	S6	S7	S8	S9	S10	S11	S12	S13	S14
Match with takepoints $TP_{s,t}$	$t2$	$t1$	$t1$	$t1$	$t1$	$t2$	$t2$	$t2$	$t2$	$t2$	$t2$	$t1$	$t1$	$t2$
Earliest fracturing day	1	1	1	1	1	39	1	273	273	273	396	379	379	1
No. of stages	57	61	54	55	64	26	97	88	86	76	63	100	100	87

The parameters and the cost coefficients are given in Table 2 while the information regarding the average flowback water and total dissolved solids (TDS) profile for a given well pad in the first 14–20 days after well pad fracturing, and the expected gas production for each well pad are obtained from Yang et al. [4].

Table 2. Parameters and cost coefficients.

Parameter	Value
Crew transition time (day)	5
Volume of fracturing fluid used per stage (m^3)	950
Freshwater used (%)	85
Storage cost ($\$/m^3$)	0.59
Freshwater trucking cost ($\$/m^3$)	29.35
Freshwater pumping cost ($\$/m^3$)	15.93
Disposal cost ($\$/m^3$)	134.18
Wastewater pumping cost ($\$/km/m^3$)	0.28
Wastewater storage cost ($\$/m^3$)	0.59
Temperature-independent base value of membrane permeability B_{WB} ($kg/(m^2\ s\ pa\ K^{1.334})$)	3.9×10^{-10}
Membrane thickness (mm)	0.65
Membrane life time (year)	4
Annual operation time (h)	8000
Heating cost ($\$/(10^9\ J)$)	5
Supply temperature (K)	293
Specific heat capacity (kJ/(kg K))	4
Average TDS concentration of the feed into membrane distillation (MD) (mg/L)	200,000

The resulting model was implemented in GAMS and solved using the general purpose global optimisation solver (BARON), which uses a branch-and-reduce algorithm to obtain a solution. Although BARON is not always guaranteed to converge to the global optimum, it has a proven track record in solving non-convex MINLP problems. The performance of BARON and statistics in solving a wide variety of test problems have been reported in the literature [30–32]. The solution comparison and the computational statistics between the three scenarios are given in Tables 3 and 4, respectively. The total volume of water required for the 14 well pads is found to be 818,800 m^3. The results encourage the use of freshwater from interruptible sources, which is achieved through piping, thereby reducing the high cost and environmental issues that are associated with trucking. It should be noted that Scenario 1, which involved the use of freshwater only, does not take into account the extra cost associated with the water network such as the cost of treatment and storage. Thus, no comparison with regard to profit is conducted between the three scenarios, as shown in Table 3. The total revenue for both Scenarios 2 and 3 is found to be $261.24 million and the total profit for Scenario 3 is found to be 0.6% higher than the profit obtained in Scenario 2. This is mainly due to the fact that the costs of wastewater disposal and treatment cost are higher in Scenario 2 compared to Scenario 3.

The fracturing schedules for the three scenarios are presented in Figures 6–8. As can be seen from these figures, the schedules involved different timing and sequences. As the schedule in Figure 6 only consider freshwater usage, the well pads fracturing followed each other depending on the availability of each well pad and also on the water availability in the impoundment. The gap between S7 and S8 in Figure 6 is due to the fact that S7 is available from day 1 while S8 only becomes available after day 273. In Figures 7 and 8, it is observed that well pad 6 is fractured last in both schedules. This is because well pad 6 has the least number of stages, which implies that it will require the lowest volume of water for fracturing, thereby reducing the volume of wastewater to be disposed in the last time point. The gaps between S8 and S9 in Figure 6, S5 and S4 in Figure 7, and S3 and S5 in Figure 8 may be attributed to what is referred to as a frac holiday, which depends mainly on water availability for fracturing.

According to the literature [4], fracturing idle time (holiday) is a flexible period when the fracturing crew takes time off, usually due to low water availability. Figures 7 and 8 show that the tightness in the fracturing schedule of each group of well pads which is much more profound in Figure 8, improve the effectiveness of flowback water reuse.

As a result of effective flowback water reuse, a saving of 183,534.65 m^3 of freshwater is achieved out of the total volume of 818,800 m^3 required for the 14 well pads. The saving is found to be 21.23% higher than those of a previous study in literature [4] that uses discrete time formulation. In Scenario 2, 96.7% of the flowback water is sent to the regenerator (R) and the remaining 3.3% is sent to the injection well to be disposed, while in Scenario 3, 99.4% of the flowback water is sent to the regenerator (R) while the remaining 0.6% is disposed.

Table 3. Solution comparison.

	Scenario 1	Scenario 2	Scenario 3
Freshwater pumped (1000, m^3)	818.80	640.30	635.30
Freshwater trucked (1000, m^3)	0	0	0
Regenerated water (1000, m^3)	0	178.53	183.53
Freshwater saved (%)	0	21.80	22.42
Freshwater trucking cost ($1000)	0	0	0
Freshwater pumping cost ($1000)	13,043	10,019	10,012
Disposal cost ($1000)	0	2119	1450
Wastewater pumping cost ($1000)	0	10.01	11.65
Wastewater storage cost ($1000)	0	1740	1747
Treatment cost ($1000)	0	11,307	10,575
Revenue ($1000)	-	261,240	261,240
Profit ($1000)	-	235,860	237,340

Table 4. Computational statistics.

	Scenario 1	Scenario 2	Scenario 3
No. of constraints	5698	9324	9418
No. of continuous variables	3796	6023	6103
No. of binary variables	210	435	435
Non-linear terms	-	1458	1514
CPU time (s)	0.11	51.82	458.59
No. of slots	14	14	14
No. of time points	15	15	15

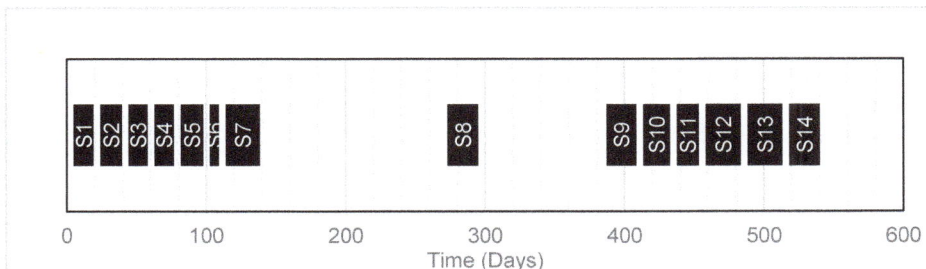

Figure 6. Fracturing schedule (base case).

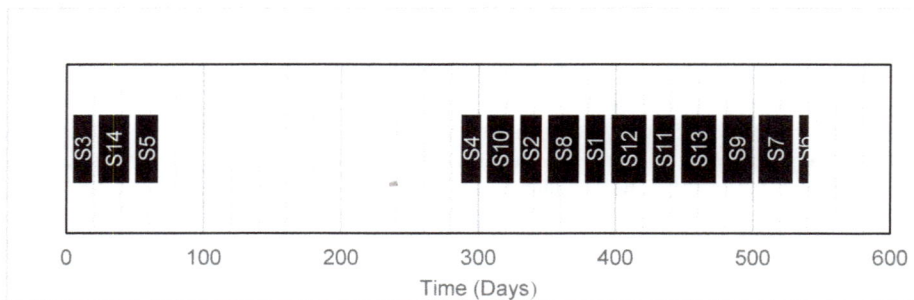

Figure 7. Fracturing schedule (Scenario 2).

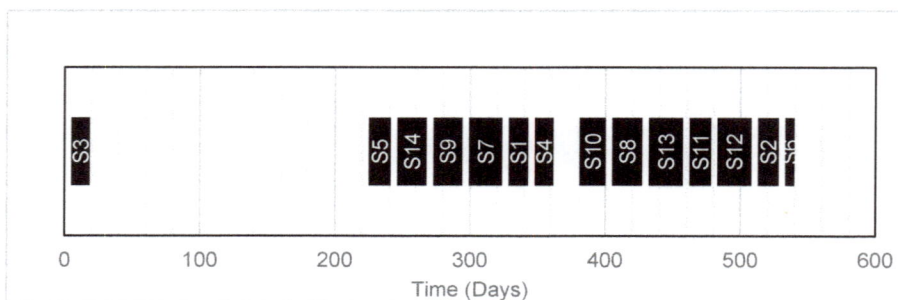

Figure 8. Fracturing schedule (Scenario 3).

In order to calculate the cost and energy associated with wastewater regeneration, cost analyses based on the black box model, standalone model, and detailed model were performed. The costs of regeneration were found to be $11.2 million, $9.8 million, and $10.5 million, respectively. The results show that the deviation of the cost function (black box model) from the actual cost of regeneration (standalone model) is 12.7%. The result obtained in Scenario 3 shows that the optimised cost of regeneration is 6.6% higher than the cost of MD standalone model. This is because optimising the temperature of the feed into MD results in a reduction of the water flux, thereby increasing the membrane area required which in turn leads to an increase in the fixed cost of the membrane. When water minimisation alone is considered, the membrane operates at the maximum feed temperature of 363 K which leads to the maximisation of the water flux across the membrane, hence the membrane area and the fixed cost are minimised. However, this does not necessarily indicate that the membrane performance is optimal, which is in agreement with the work of Elsayed et al. [9]. The design specifications for the optimal design of the MD regenerator are given in Table 5. The optimal feed temperature was found to be 354 K and the membrane area required was 186.67×10^3 m^2. The permeate flux, thermal efficiency, thermal energy required, and the removal ratio are also given in Table 5. The model prediction of 0.013 kg/(m^2 s) fow Much Water Does U.S [9], as well as the experimental data of 0.0125 kg/(m^2 s) at 351 K reported by Yun et al. [20].

The simultaneous optimisation of both energy and water within the water network results in a 12.7% reduction in the amount of energy required by the regenerator based on the throughput per day. The amount of energy required is reduced from 699×10^6 kJ (equivalent to 18,250 m^3 of natural gas) to 610×10^6 kJ (equivalent to 15,926 m^3 of natural gas). The value of energy consumed by the regenerator is 244×10^3 kJ/m^3 of distillate, which is found to be less than the range of thermal energy reported in the literature for membrane distillation. The range of thermal energy required by membrane distillation is between 120 and 1700 kWh/m^3, equivalent to between 432×10^3 kJ/m^3 and

6.12×10^6 kJ/m^3 [23,33]. The average volume of flared gas per unit time based on literature [18] is used in this study and this is compared to the energy requirement of the regenerator. Gas that would otherwise be flared is used as the source of heat for the regenerator, thereby, making the heating cost in the objective function to become zero.

Table 5. Design specification for MD.

Design Variables	Optimum Values
MD feed temperature (K)	354
Required membrane area (m^2)	186.67×10^3
Thermal efficiency	0.98
Thermal energy (kJ/day)	610×10^6
Permeate flux (kg/(m^2 s))	0.013
Removal ratio (RR) (%)	1

6. Conclusions and Recommendations for Future Work

This work explores simultaneous water and energy optimisation in shale play using continuous time formulation with the incorporation of a detailed MD model within the water network. The goal is to balance the trade-off between water acquisition from interruptible and uninterruptible water sources. It was shown that water acquisition from interruptible sources through piping can lead to a reduction in both the freshwater cost and the high environmental impact associated with trucking water from uninterruptible sources. The results also demonstrated that for the considered case study, membrane distillation is capable of handling wastewater from hydraulic fracturing effectively, so that 99.4% of the flowback water generated is treated and reused. The efficient reuse of wastewater leads to a 22.42% reduction in the amount of freshwater required. The importance of simultaneously optimising the fracturing schedule with water and energy management was demonstrated. The approach indicates that optimising energy and water simultaneously results in a significant reduction in the amount of thermal energy required for regeneration. It is difficult to find the specific amount/volume of gas flared per well pad in the literature. However, based on the average data gathered from the literature, the amount of gas that is flared in most shale play is sufficient to provide the energy needed for regeneration. Considering the uncertainties associated with shale gas exploration in terms of water usage for hydraulic fracturing, flowback water generation, the cost associated with water management, and the price of oil and gas, future work will address the uncertainties associated with the process and the possible impact of such uncertainties on the overall project. Future work can also consider multiple desalination technologies and the integration of fossil energy with renewable sources to reduce the carbon footprint of the resulting network [34,35]. Hence, sustainability-based objective functions can be used to optimise the system design [36,37].

Author Contributions: Conceptualisation, D.O., T.M., R.M., D.S., and M.E.-H.; Data curation, D.O. and T.M.; Formal analysis, D.O., T.M., R.M., and D.S.; Funding acquisition, T.M.; Investigation, D.O. and T.M.; Methodology, D.O., T.M., and M.E.-H.; Resources, T.M.; Software, D.O., T.M., R.M., D.S., and M.E.-H.; Supervision, T.M. and M.E.-H.; Validation, D.O., T.M., R.M., D.S., and M.E.-H.; Visualisation, D.O., T.M., R.M., D.S., and M.E.-H.; Writing–original draft, D.O.; Writing–review and editing, D.O., T.M., R.M., D.S., and M.E.-H.

Funding: This research was funded by the National Research Foundation (NRF), South Africa.

Conflicts of Interest: The authors declare no conflict of interest.

Nomenclature

Sets

D	$\{d \mid d = \text{injection well}\}$
N	$\{n \mid n = \text{time point}\}$
S	$\{s \mid s = \text{well pad}\}$
T	$\{t \mid t = \text{an interruptible source and its corresponding impoundment}\}$

$TP_{s,t}$	Match between well pad s and source t
Y	$\{y \mid y = \text{historical river flowrate data year}\}$
Parameters	
AT_s	Availability time of well pad s, day
AOT	Annual operating time, h
B_{wb}	Temperature independent base value for the permeability, $\text{kg}/\text{m}^2.\text{s.pa.K}^{1.334}$
C_p	Specific heat capacity of the feed stream, $\text{KJ}/(\text{kg K})$
C^{max}	Maximum inlet concentration in the treatment unit, $\text{mg}/\text{L } t$
Cf^{feed}	Concentration of the feed water in MD, mg/L
CS^{max}	Maximum inlet concentration in well pad s, mg/L
CS_s	Flowback water concentration in well pad s, mg/L
$CTS_S^{s'}$	Crew transition time between well pads, day
DI^{max}	Maximum capacity of injection well d, m^3
DS_s	Distance from well pad s to a treatment facility, km
H	Time horizon of interest, day
LR	Liquid recovery for the regenerator
NY	Number of historical year, year
$OC_s^{pump,fw}$	Freshwater pumping cost, $\$/\text{m}^3$
$OC_s^{truck,fw}$	Freshwater trucking cost, $\$/\text{m}^3$
$OC_s^{pump,ww}$	Wastewater pumping cost, $\$/\text{m}^3/\text{km}$
OC^{dis}	Cost of wastewater disposal, $\$/\text{m}^3$
$OC^{st,ww}$	Cost of wastewater storage, $\$/\text{m}^3$
OC^{ht}	Cost of heating, $\$/(10^9 \text{ J})$
P_s	Gas production of well pad s, m^3
SP^{gas}	Unit price of natural gas, $\$/\text{m}^3$
ST_s	Availability date of well pad s, day
TR_s	Time required fracturing well pad s, day
T_{sf}	Temperature of feed water in the treatment unit, K
u	Ratio of recycled reject to raw feed
V^{max}	Maximum capacity of storage, m^3
V^{min}	Minimum capacity of storage, m^3
WR_s	Amount of water required to fracture well pad s, m^3
X_{NaCl}	Molar concentration of NaCl in the feed
δ	Membrane thickness, mm
∂_{ED}	Energy density, kJ/m^3
ρ_{water}	Density of water, kg/m^3
Binary variables	
$w_{s,n}$	Defines the beginning of stimulating each well pad s at time point n
$wv_{s,n}$	Transfer of water from well pad s to storage at time point n
wr_n	Transfer of water from storage to the regenerator at time point n
Continuous variables	
A_m	Required membrane area, m^2
AFC	Annualised fixed capital cost for the regenerator, $\$/\text{year}$
AHC	Annualised heating cost for the regenerator, $\$/\text{year}$
AOC	Annualised operating cost for the regenerator, $\$/\text{year}$
Bw	Membrane permeability, $\text{kg}/(\text{m}^2 \text{ pa})$
$c_{s,n}^{fbw}$	Flowback water concentration of well pad s at time point n, mg/L
$c_n^{st,ww}$	Contaminant concentration in the treatment unit at time point n, mg/L
c_n^{perm}	Outlet concentration of contaminant from the regenerator at time point n, mg/L
c_n^{con}	Contaminant concentration removed from the water by the regenerator at time point n, mg/L
cp^{perm}	Permeate concentration from MD, mg/L
cr^{con}	Retentate concentration from MD, mg/L
$du_{s,n}$	Duration of well pad s at time point n, day
E^{cons}	Thermal energy consumption per unit of water treated, kJ/m^3

E_n^{total}	Thermal energy required at time point n, kJ
$f_{s,n}$	Total water required to fracture well pad s at time point n, m^3
$f_{t,n,y}^{pump}$	Water pumped from interruptible source at time point n in scenario year y, m^3
$f_{t,n,y}^{truck}$	Water trucked from uninterruptible source at time point n in scenario year y, m^3
$f_{s,n}^{fw}$	Freshwater required to fracture well pad s at time point n, m^3
$f_{s,n}^{ww}$	Wastewater required to fracture well pad s at time point n, m^3
$f_{s,n}^{fbw}$	Flowback water from well pad s at time point n, m^3
$f_{s,n}^{st}$	Flowback water sent to storage tank from well pad s at time point n, m^3
$f_{s,n}^{dis}$	Flowback water sent to disposal from well pad s at time point n, m^3
f_n^{reg}	Total flowback water to be treated at time point n, m^3
f_n^{perm}	Amount of water collected as permeate from the regenerator at time point n, m^3
f_n^{con}	Amount of retentate from the regenerator at time point n, m^3
fd_n	Total water sent to disposal at time point n, m^3
$ff_{d,n}^{dis}$	Throughput of an injection well d at time point n, m^3
ff^{MD}	Total flowrate into MD, m^3/day
ff^{feed}	Total flowrate into MD, kg/day
ff^{perm}	Permeate flowrate from MD, kg/day
ff^{con}	Retentate flowrate from MD, kg/day
$i_{t,n}^{fw}$	Total freshwater required from impoundment t for fracturing at time point n, m^3
Jw	Water flux across the membrane, kg/(m^2·s)
k_m	Membrane thermal conductivity, kW/(m·K)
$p_{s,n}$	Expected gas production of well pad s at time point n, m^3
p_{wf}^{vap}	Water vapour pressure of the feed in MD, pa
p_{wp}^{vap}	Water vapour pressure of the permeate in MD, pa
Q	Heat required by the feed into MD, kJ/day
RR	Regenerator removal ratio
$ts_{s,n}$	Start time of well pad s at time point n, day
$tf_{s,n}$	Finish time of well pad s at time point n, day
tt_n	Time that corresponds to time point n, day
tr_n	Start time of regeneration at time point n, day
ttr_n	Duration of regeneration at time point n, day
$tv_{s,n}$	Time at which water is transferred from well pad s to storage tank at time point n, day
T_{mf}	Temperature of the feed on the membrane, K
T_{mp}	Temperature of the permeate on the membrane, K
T_m	Membrane average temperature, K
T_{bf}	Temperature of the feed in the bulk, K
T_{bp}	Temperature of permeate in the bulk, K
$vi_{t,n,y}$	Volume of impoundment t at time point n in scenario year y, m^3
v_n^{ww}	Capacity of wastewater tank at time point n, m^3
$v_{s,n}^{ft}$	Capacity of fracturing tank on well pad s at time point n, m^3
v_n^{nat}	Volume of natural gas needed to produce the required energy at time point n, m^3
x_{wf}	Mole fraction of water in the feed
γ_{wf}	Activity coefficient of water in the feed for membrane distillation
η	Overall thermal efficiency of the regenerator
ΔH_{vw}	Latent heat of vaporisation for water, kJ/kg
θ	Temperature polarisation coefficient

Superscript

con	Concentrate
cons	Consumption
dis	Disposal
feed	Feed
ft	Fracturing tank
fw	Freshwater
fbw	Flowback water

gas	Gas
ht	Heating
max	Maximum
min	Minimum
nat	Natural gas
pump	Pumping
perm	Permeate
reg	Regenerator
st	Storage
total	Total
truck	Trucking
vap	Vapour
ww	Wastewater
Subscript	
bp	Permeate bulk
bf	Feed bulk
m	Membrane
mp	Membrane permeate
mf	Membrane feed
wf	Feed water
wp	Permeate water

References

1. Al-Douri, A.; Sengupta, D.; El-Halwagi, M.M. Shale Gas Monetization—A review of downstream processing of chemical and fuels. *J. Nat. Gas. Eng.* **2017**, *45*, 436–455. [CrossRef]
2. Zhang, C.; El-Halwagi, M.M. Estimate the capital cost of shale-gas monetization projects. *Chem. Eng. Prog.* **2017**, *113*, 28–32.
3. Arthur, J.D.; Bohm, B.; Layne, M. Hydraulic fracturing considerations for natural gas wells of the Marcellus Shale. In Proceedings of the Ground Water Protection Council 2008 Annual Forum, Cincinnati, OH, USA, 21–24 September 2008; pp. 1–16.
4. Yang, L.; Grossmann, I.E.; Manno, J. Optimization models for shale gas water management. *AIChE J.* **2014**, *60*, 3490–3501. [CrossRef]
5. Vengosh, A. How Much Water Does U.S. Fracking Really Use? *2015*. Available online: https://today.duke.edu/2015/09/frackfoot (accessed on 25 May 2018).
6. Hasaneen, R.; El-Halwagi, M.M. Integrated process and microeconomic analyses to enable effective environmental policy for shale gas in the United States. *Clean Technol. Environ. Policy* **2017**, *19*, 1775–1789. [CrossRef]
7. Petrakis, S. Reduce cooling water consumption: New closed loop cooling method improves process cooling tower operations. *Hydrocarb. Process.* **2008**, *87*, 95–98.
8. Elsayed, N.A.; Barrufet, M.A.; Eljack, F.T.; El-Halwagi, M.M. Optimal design of thermal membrane distillation systems for the treatment of shale gas flowback water. *Int. J. Membr. Sci.* **2015**, *2*, 1–9.
9. Elsayed, N.A.; Barrufet, M.A.; El-Halwagi, M.M. Integration of thermal membrane distillation networks with processing facilities. *Ind. Eng. Chem. Res.* **2013**, *53*, 5284–5298. [CrossRef]
10. Yang, L.; Grossmann, I.E.; Mauter, M.S.; Dilmore, R.M. Investment optimization model for freshwater acquisition and wastewater handling in shale gas production. *AIChE J.* **2015**, *61*, 1770–1782. [CrossRef]
11. Gao, J.; You, F. Optimal design and operations of supply chain networks for water management in shale gas production: MILFP model and algorithm for the water-enegy nexus. *AIChE J.* **2015**, *61*, 1184–1208. [CrossRef]
12. Gao, J.; You, F. Deciphering and handling uncertainty in shale gas supply chain design and optimization: Novel modelling framework and computational efficient solution algorithm. *AIChE J.* **2015**, *61*, 3739–3760. [CrossRef]
13. Bartholomew, T.V.; Mauter, M.S. Multiobjective optimization model for minimizing cost and environmental impact in shale gas water and wastewater management. *ACS Sustain. Chem. Eng.* **2016**, *4*, 3728–3735. [CrossRef]
14. Lira-Barragán, L.; Ponce-Ortega, J.M.; Guillén-Gosálbez, G.; El-Halwagi, M.M. Optimal Water Management under Uncertainty for Shale Gas Production. *Ind. Eng. Chem. Res.* **2016**, *55*, 1322–1335. [CrossRef]

15. Yang, L.; Salcedo-Diaz, R.; Grossmann, I.E. Water network optimization with wastewater regeneration models. *Ind. Eng. Chem. Res.* **2014**, *53*, 17680–17695. [CrossRef]

16. Kamrava, S.; Gabriel, K.J.; El-Halwagi, M.M.; Eljack, F.T. Managing abnormal operation through process integration and cogeneration systems. *Clean Techn. Environ. Policy* **2015**, *17*, 119–128. [CrossRef]

17. Kazi, M.K.; Eljack, F.; Elsayed, N.A.; El-Halwagi, M.M. Integration of energy and wastewater treatment alternatives with process facilities to manage industrial flares during normal and abnormal operations: A multi-objective extendible optimization framework. *Ind. Eng. Chem. Res.* **2016**, *55*, 2020–2034. [CrossRef]

18. Glazer, Y.R.; Kjellsson, J.B.; Sanders, K.T.; Webber, M.E. Potential for using energy from flared gas for on-site hydraulic fracturing wastewater treatment in Texas. *Environ. Sci. Technol. Lett.* **2014**, *1*, 300–304. [CrossRef]

19. Wikramanayake, E.; Bahadur, V. Flared natural gas-based onsite atmospheric water harvesting (AWH) for oilfield operations. *Environ. Res. Lett.* **2016**, *11*, 1748–9326. [CrossRef]

20. Yun, Y.; Ma, R.; Zhang, W.; Fane, A.G.; Li, J. Direct contact membrane distillation mechanism for high concentration NaCl solutions. *Desalination* **2006**, *188*, 251–262. [CrossRef]

21. Gregory, K.B.; Vidic, R.D.; Dzombak, D.A. Water management challenges associated with the production of shale gas by hydraulic fracturing. *Elements* **2011**, *7*, 181–186. [CrossRef]

22. Lawson, K.W.; Lloyd, D.R. Membrane distillation: Review. *J. Membr. Sci.* **1997**, *124*, 9–25. [CrossRef]

23. Camacho, L.M.; Dume´e, L.; Zhang, J.; Li, J.D.; Duke, M.; Gomez, J.; Gray, S. Advances in membrane distillation for water desalination and purification applications. *Water* **2013**, *5*, 94–196. [CrossRef]

24. Estrada, J.M.; Bhamidimarri, R. A review of the issues and treatment options for wastewater from shale gas extraction by hydraulic fracturing. *Fuel* **2016**, *182*, 292–303. [CrossRef]

25. Engle, M.A.; Rowan, E.L. Geochemical evolution of produced waters from hydraulic fracturing of the Marcellus Shale, northern Appalachian Basin: A multivariate compositional data analysis approach. *Int. J. Coal Geol.* **2014**, *126*, 45–56. [CrossRef]

26. El-Halwagi, M.M. *Sustainable Design through Process Integration: Fundamentals and Applications to Industrial Pollution Prevention, Resource Conservation, and Profitability Enhancement*; Elsevier: Oxford, UK, 2012.

27. Khayet, M.; Matsuura, T. *Membrane Dstillation: Principles and Applications*; Elsevier: Amsterdam, The Netherlands, 2011.

28. Schofield, R.W.; Fane, A.G.; Fell, C.J.D. Heat and mass transfer in membrane distillation. *J. Membr. Sci.* **1987**, *33*, 299–313. [CrossRef]

29. Al-Obaidani, S.; Curcio, E.; Macedonio, F.; Di Profio, G.; Al-Hinai, H.; Drioli, E. Potential of membrane distillation in seawater desalination: thermal efficiency, sensitivity study and cost estimation. *J. Memb. Sci.* **2008**, *323*, 85–98. [CrossRef]

30. MINLPLib. Available online: http://www.minlplib.org (accessed on 19 June 2018).

31. Nohra, C.J.; Sahinidis, N.V. Global optimization of nonconvex problems with convex-transformable intermediates. *J. Glob. Optim.* **2018**, 1–22. [CrossRef]

32. Tawarmalani, M.; Sahinidis, N.V. A polyhedral branch-and-cut approach to global optimization. *Math. Program.* **2005**, *103*, 225–249. [CrossRef]

33. Suarez, F.; Urtubia, R. Tackling the water-energy nexus: an assessment of membrane distillation driven by salt-gradient solar ponds. *Clean Technol. Environ. Policy* **2016**, *18*, 1–16. [CrossRef]

34. Al-Aboosi, F.Y.; El-Halwagi, M.M. An integrated approach to water-energy nexus in shale gas production. *Processes* **2018**, *6*, 52. [CrossRef]

35. Baaqeel, H.; El-Halwagi, M.M. Optimal multi-scale capacity planning in seawater desalination systems. *Processes* **2018**, *6*, 68. [CrossRef]

36. El-Halwagi, M.M. A return on investment metric for incorporating sustainability in process integration and improvement projects. *Clean Technol. Environ. Policy* **2017**, *19*, 611–617. [CrossRef]

37. Guillen-Cuevas, K.; Ortiz-Espinoza, A.P.; Ozinan, E.; Jiménez-Gutiérrez, A.; Kazantzis, N.K.; El-Halwagi, M.M. Incorporation of safety and sustainability in conceptual design via a return on investment metric. *ACS Sustain. Chem. Eng.* **2018**, *6*, 1411–1416. [CrossRef]

![processes logo] *processes*

MDPI

Article

Hybrid Approach for Optimisation and Analysis of Palm Oil Mill

Steve Z. Y. Foong [1], Viknesh Andiappan [2]🄳, Raymond R. Tan [3], Dominic C. Y. Foo [1]🄳 and Denny K. S. Ng [1,2,]*🄳

[1] Department of Chemical and Environmental Engineering, The University of Nottingham Malaysia Campus, Broga Road, 43500 Semenyih, Malaysia; kebx6fzy@exmail.nottingham.edu.my (S.Z.Y.F.); Dominic.Foo@nottingham.edu.my (D.C.Y.F.)
[2] School of Engineering and Physical Sciences, Heriot-Watt University Malaysia, 62200 Putrajaya, Wilayah Persekutuan Putrajaya, Malaysia; v.murugappan@hw.ac.uk
[3] Centre for Engineering and Sustainable Development Research, De La Salle University, 2401 Taft Avenue, 0922 Manila, Philippines; raymond.tan@dlsu.edu.ph
* Correspondence: Denny.Ng@hw.ac.uk; Tel.: +60-3-8894-3784

Received: 11 January 2019; Accepted: 11 February 2019; Published: 15 February 2019

Abstract: A palm oil mill produces crude palm oil, crude palm kernel oil and other biomass from fresh fruit bunches. Although the milling process is well established in the industry, insufficient research and development reported in optimising and analysing the operations of a palm oil mill. The performance of a palm oil mill (e.g., costs, utilisation and flexibility) is affected by factors such as operating time, capacity and fruit availability. This paper presents a hybrid combined mathematical programming and graphical approach to solve and analyse a palm oil mill case study in Malaysia. The hybrid approach consists of two main steps: (1) optimising a palm oil milling process to achieve maximum economic performance via input-output optimisation model (IOM); and (2) performing a *feasible operating range analysis* (FORA) to study the utilisation and flexibility of the developed design. Based on the optimised results, the total equipment units needed is reduced from 39 to 26 unit, bringing down the total capital investment by US$6.86 million (from 18.42 to 11.56 million US$) with 23% increment in economic performance (US$0.82 million/y) achieved. An analysis is presented to show the changes in utilisation and flexibility of the mill against capital investment. During the peak crop season, the utilisation index increases from 0.6 to 0.95 while the flexibility index decreases from 0.4 to 0.05.

Keywords: mathematical programming; graphical approach; feasible operating range analysis; utilisation index; flexibility index

---

1. Introduction

Oil palm (*Elaeis guineensis*) is cultivated for the production of fresh fruit bunches (FFBs) due to its stability, high yield and low cost [1,2]. FFBs are then can be converted into a variety of products including foods, cosmetics, detergents and biofuels. To date, approximately 85% of global crude palm oil (CPO) is produced in Indonesia and Malaysia [3]. CPO is extracted from FFBs in processing facilities known as palm oil mills (POM). A typical milling process consists of several operational units as shown in Figure 1. FFBs undergo sterilisation, threshing, digestion and pressing to produce pressed liquid and cake. The pressed liquid is clarified and purified to produce CPO, while the pressed cake undergoes nut separation, nut cracking, kernel separation and drying to produce palm kernel (PK). Most POMs in Malaysia will send the PK to a kernel crushing plant for crude palm kernel oil (CPKO) production [4] before refinery processes where CPO and CPKO are refined into higher quality edible oils and fats [5]. Throughout the milling process, biomass such as palm kernel shell (PKS), pressed empty fruit bunch (PEFB) and palm pressed fibre (PPF) are generated as by-products. Meanwhile, large amounts of strong wastewater, which is known as palm oil mill effluent (POME) are produced during sterilisation and clarification operations.

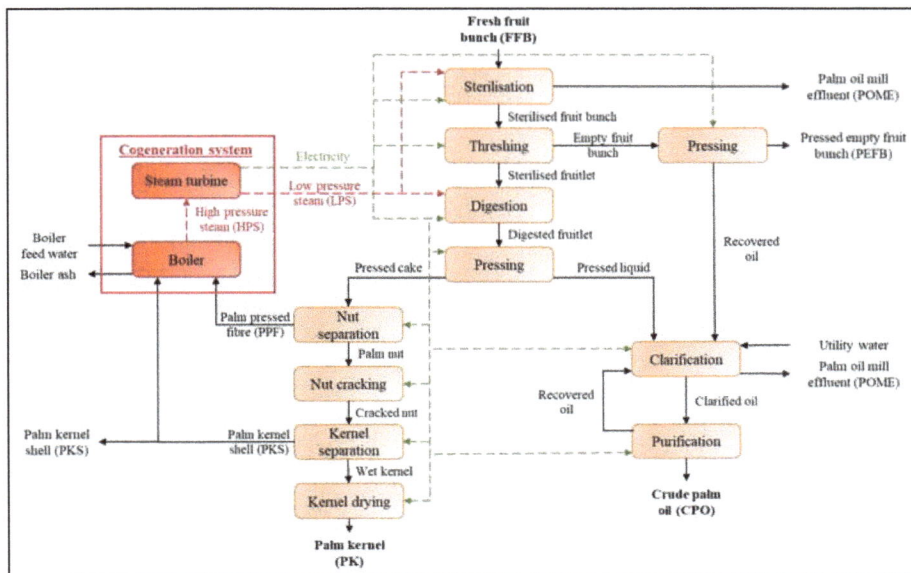

Figure 1. Unit operations in a typical palm oil mills (POM) [6].

POMs are usually located near to the plantations, which usually are in remote areas to minimise logistics costs. In Malaysia, 63% of the active POMs are positioned far away (>10 km) from electrical grid connection point [7], leaving them at a disadvantage as they would require steam and electricity for CPO extraction. Abdullah and Sulaiman [8] estimated that 0.075–0.1MWh electricity and 2.5 t of low-pressure steam (LPS) are required per ton of CPO produced. In current practice, over three-quarters of over 400 POMs in Malaysia met the process steam and electricity demands by burning PPF and a portion of PKS generated from the milling process [9,10] via co-generation [11]. Excess PKS can then be sold as an alternative solid fuel around the world [12,13], while PEFB is returned to plantations as mulching materials [14] or composted to produce biofertilizer [15]. The biomass can also be used for a range of other applications (e.g., pellet, dried long fibre, etc.). Meanwhile, pond-based wastewater treatment systems are commonly used to treat POME before discharge [16].

Yu-Lee [17] stated that the processing capacity of a plant or system depends on the labour, equipment, technology and materials available. In this sense, POMs would have their unique design features and the operations of each mill may differ between one another. For instance, the capacity of a typical POM could range between 20 to 90 t/h of FFB, with operations up to 19 h every day [18]. Besides, ripe FFBs collected from plantations must be transported and processed immediately in POMs to prevent degradation of CPO quality due to increased free fatty acid content [19]. The amount of FFBs supplied to a POM could vary depending on location and time, due to seasonal crop changes and possible unforeseen circumstances in the plantations [6,20]. To overcome these issues, most plants or systems including POM are often built with an excess capacity to ensure higher flexibility [21] and lower processing costs (i.e., labour, service and maintenance costs) [22]. However, this affects the utilisation and economic performance of POM, especially during the lean crop season.

According to the literature, there are several methods developed to optimise and analyse the performance of systems; one of the commonly used methods is input-output (IO) model. IO model was first developed by Leontief [23] to deal with the interdependencies between system components (e.g., materials, processes, costs) using systems of linear equations. IO models are used to study the behaviour of a system when the input or output of one system component changes quantitatively [24]. Some notable works on IO model have been presented to analyse economic networks [23], industrial networks [25], chemical industry supply chains [26], food manufacturing plants [27] and life cycle assessment [28,29]. IO optimisation models (IOM) have also been developed based on the general IO methodology. IOM has been successfully applied for industrial complexes [30,31], biorefineries [32], sustainable industrial systems [33], human resources [34] and palm oil plantations [35] to make the best use of situation, goods or production capacity.

Apart from IOM, graphical approaches have been developed to analyse system performance. Graphical approaches provide visual assistance in analysing scientific data and communicating quantitative information [36]. Some of the well-known graphical approaches are the insight-based *pinch analysis technique* [37] and *process graph*, also known as P-graph [38]. Detailed information and applications of such approaches have been reviewed and discussed by Linnhoff [39], Foo [40], and Teng et al. [41]. Recently, Andiappan et al. [42] proposed the *feasible operating range analysis* (FORA) to examine the real-time feasible operating range of an energy system graphically. Such approach allows the range output (i.e., maximum and minimum of each output) of a system to be determined, considering material input and capacity constraints of individual unit operations. Besides, it also provides insight into potential design modifications based on variations in output demand and process bottleneck [43].

The studies presented thus far provide evidence for the applications of mathematical programming and graphical approaches (i.e., IOM and FORA) to optimise and analyse problems in various fields. However, limited works were reported for a hybrid approach to deal with such issues. None of the contributions discussed has focused on palm oil milling processes apart from Foong et al. [6], in which a mathematical programming approach alone is presented. Based on the previous work [6], operational variables such as operating hours and labour costs are yet to be considered. Besides, analysis on a real-time feasible operating range and the bottleneck of the developed design is not performed in the previous work. In addition, the operational performance of the milling process can be quantified in terms of utilisation and flexibility indices, introduced by Grossmann et al. [44] to measure the usage and expected deviation from a nominal design state that a process can handle. These research gaps are dealt with in this study, developing a hybrid approach consisting of IOM for palm oil mill optimisation, followed by FORA to analyse the feasible operating range of the developed system. In particular, this work provides an extended account of FORA, whereby production rates, flexibility and utilisation indices and capital expenditure are considered simultaneously to provide a visualisation tool for process improvement.

In the following section, the problem statement for this work is presented, followed by a detailed formulation for IOM in Section 3. Next, an existing POM flowsheet is optimised using the input-output approach described in Section 4. Following this, the economic performance, utilisation and flexibility of the POM are then compared to highlight the improvements achieved. Lastly, the conclusions and prospective future works are described in the final section.

2. Problem Statement

The problem addressed by the proposed approach is divided into two parts, stated as follows. The palm oil milling processes consist of a set of technology $te \in TE$ with interchangeable material $m \in M$. Firstly, an IOM is developed where $\mathbf{A}$ is the input and output matrix composed of the fixed interaction ratios, $a_{m,te}$ between material m and technology te. Each crop season s has a fixed fraction of occurrence, α_s, to indicate the proportion of each year that it takes up. Different levels of supply of material m are available in each crop season s. The number of equipment units operated, $\mathbf{U}_{te}$ determined from the nominal capacity, $\mathbf{CAP}_{te}$ available in the market. Each material m and technology te associated with a given material cost, C_m, operating cost, OC_{te}, capital cost, CC_{te} and electricity consumption, $\mathbf{E}_{te}$, respectively. In the event where annual operating time, AOT exceeds the annual shift time, AST, additional overtime cost, OTC and operating costs, $OPEX$ required. The objective is to maximise the economic performance, EP of the POM as shown in Equation (1).

$$\text{Maximise } EP \tag{1}$$

Based on the optimised POM design, the $\boldsymbol{U}_{te}$ determined is set as the maximum units operated, $\mathbf{U}_{te}^{\text{max}}$ to identify the technology bottleneck, $\boldsymbol{B}_{te}$ from the maximum capacity, $\mathbf{CAP}_{te}^{\text{max}}$ of each technology te. Next, FORA is then performed to evaluate the developed system using utilisation and flexibility indices, UI and FI, respectively. The following section further explains the approach developed for this work.

3. Hybrid Approach Formulation

As mentioned previously, a hybrid approach is developed in this work to optimise the palm oil milling process via IOM, followed by FORA to analyse the developed system. The italic notations represent the variables determined by the model and non-italic notations represent constant parameters defined in the proposed approach. Meanwhile, matrix and vector symbols are represented by bold notations.

3.1. Input-Output Optimisation Model (IOM)

In this model, each crop season in which material flows would vary is represented by index s. It is assumed that a linear correlation for material flows in the milling process is given in Equation (2)

$$\mathbf{A}(\boldsymbol{x}_{te})_s = (\boldsymbol{y}_m)_s \; \forall m, \qquad \forall s \tag{2}$$

where $\mathbf{A}$ is the matrix consists of fixed interaction ratios, $a_{m,te}$ for material input and output ratios, to and from technology te. Each column in matrix $\mathbf{A}$ corresponds to different technology te, while its rows correspond to material m flows. $a_{m,te}$ are expressed in negative values for material inputs, positive values for material outputs or zero if there are no interactions between material m and technology te. $\boldsymbol{x}_{te}$ is the processing capacity vector of technology te, in which positive values obtained for technologies operated and zero when it is not. Meanwhile, $\boldsymbol{y}_m$ is the flow rate vector of material m (i.e., input or output). Final and by-products are indicated with positive values while process feedstocks are indicated with negative values and intermediates denoted with zeros. Note that both $\boldsymbol{x}_{te}$ and $\boldsymbol{y}_m$ are expressed in material flow rate (t/h) or power generation (kW).

In the process, electricity is also being consumed to operate technology te for material conversions. However, electricity demand, E^{Demand} of a POM relies on the number of units operated for technology, $\boldsymbol{U}_{te}$ rather than linear correlation as shown in Equation (3).

$$\left(E^{\text{Demand}}\right)_s = \sum_{te=1}^{TE} (\boldsymbol{U}_{te})_s \mathbf{E}_{te} \qquad \forall s \tag{3}$$

$\mathbf{E}_{te}$ is a diagonal matrix for electricity consumption specified per unit technology te operated. Vector for the number of units of technology operated, $\boldsymbol{U}_{te}$ is determined based on the inverse of a nominal capacity diagonal matrix, $\mathbf{CAP}_{te}$ available in the market ($\mathbf{CAP}_{te}^{-1}$) obtained from Equation (4).

$$(\boldsymbol{U}_{te})_s \geq (\boldsymbol{x}_{te})_s \mathbf{CAP}_{te}^{-1} \; \forall s, \qquad \forall te \tag{4}$$

$\boldsymbol{U}_{te}$ consists of positive integers and the inequality in Equation (4) ensures that the products of $\boldsymbol{U}_{te}$ and $\mathbf{CAP}_{te}$ to be greater or equal to $\boldsymbol{x}_{te}$ for the process to operate.

In the presence of power supply from grid connection, the system produces and utilises electricity generated onsite. To ensure that the process is self-sufficient without interruption, an additional constraint, Equation (5) is included whereby the output of electricity produced, $y_{electricity}$ in the process is greater or equal to the electricity demand, E^{Demand} in each crop season s.

$$\left(y_{electricity}\right)_s \geq \left(E^{Demand}\right)_s \qquad \forall s \tag{5}$$

Note that the focus of this work is to model the interdependency of each equipment with one another in a single system or plant. For conservative measurement, the power consumption and process efficiency for maximum loading is assumed for each operating equipment to prevent underestimation of power demand needed, regardless of the process throughput for each equipment. Every technology unit te is sized based on these conservative values to ensure the reliability of system developed. As such, every time an equipment is selected, a conservative energy consumption value (or maximum) is activated.

Meanwhile, the economic performance, EP of the process is evaluated based on Equation (6)

$$EP = GP - CRF \times CAPEX \tag{6}$$

where GP, CRF and $CAPEX$ represent the gross profit, capital recovery factor and capital costs required, respectively. To ensure that the developed system can sustain itself economically, EP must be greater or equal to zero. Next, Equation (7) is used to calculate GP

$$GP = \sum_s \alpha_s \left[\left(AOT \sum_{m=1}^{M} y_m C_m - OPEX - OTC \right)_s - LC \right] \tag{7}$$

whereby AOT, α_s, C_m, $OPEX$, OTC and LC are the annual operational time, fraction of occurrence, material, total operating, overtime and labour costs, respectively. Equation (7) is subject to

$$\sum_s \alpha_s = 1 \tag{8}$$

in which the inclusion of α_s assessed the performance of the system developed in all crop season s. Each fraction of occurrence represents the time fraction where a season occurs. The summation of these fractions must equal to one as shown in Equation (8) as the time fraction is obtained by dividing the duration of a crop season s with the total duration considered. AOT is determined by Equation (9)

$$\left(m^{max}\right)_s \geq \left(AOT \times y_m\right)_s \qquad \forall s \tag{9}$$

where m^{max} is the maximum material demand (positive value) or available (negative value) per annum, depending on the constraint set for each season s. Equation (9) is subject to

$$\left(AOT\right)_s \leq AOT^{max} \qquad \forall s \tag{10}$$

where AOT^{max} is the maximum annual operating time of the process.

CRF is used to annualise $CAPEX$ over a specified operation lifespan t_{te}^{max} and discount rate, r, determined via Equation (11).

$$CRF = \frac{r\left(1+r\right)^{t_{te}^{max}}}{\left(1+r\right)^{t_{te}^{max}} - 1} \tag{11}$$

CAPEX is calculated based on the units of technology installed during the high crop season, $\left(U_{te}\right)_H$ while $OPEX$ depends on the units of technology operated, U_{te} in the process as shown in Equations (12)–(13).

$$CAPEX = \sum_{te=1}^{TE} \left(U_{te}\right)_H CC_{te} \tag{12}$$

$$(OPEX)_s = \sum_{te=1}^{TE} (U_{te})_s OC_{te} \quad \forall s \tag{13}$$

CC_{te} and OC_{te} represent the capital and operating costs per unit of technology te, expressed in diagonal matrixes. Meanwhile, Equations (14) and (15) determine OTC and LC required.

$$(OTC)_s = C_{OT} n_{wk} [(AOT)_s - AST] \qquad \forall s \tag{14}$$

$$LC = C_{lab} n_{wk} n_{ws} \tag{15}$$

where C_{OT} and C_{lab} are the specific overtime cost and labour cost; n_{wk} and n_{ws} represent the number of workers and working shifts per day; AST is the annual shift time of the process.

3.2. Feasible Operating Range Analysis (FORA)

It is worth mentioning that the optimal design obtained using IOM is only optimised for a given set of conditions. When changes arise in the near future, it is important to have sufficient flexibility to cater for such changes. As such, FORA provide a clear visualisation to avoid the system developed from over- or under-designed. In fact, it provides flexibility for the decision maker to decide on the required design flexibility based on how much *CAPEX* to be invested. Based on the IOM developed previously, FORA is performed to analyse the feasible operating range of the POM designed. The analysis begins by setting the maximum units of technology installed, U_{te}^{max} as the U_{te} of the design with the smallest capacity (i.e., during low crop season) as given in Equation (16).

$$U_{te}^{max} = (U_{te})_L \tag{16}$$

The product of U_{te}^{max} and CAP_{te} gives the maximum capacity, CAP_{te}^{max} as shown in Equation (17) and the inverse matrix, $(CAP_{te}^{max})^{-1}$ is used in Equation (18) to identify the technology bottleneck, B_{te} of the system. B_{te} ranges from zero to one where zero indicating that technology te is not utilised, while one shows the bottleneck of the entire system in which the capacity of that particular technology te is utilised to its maximum potential.

$$CAP_{te}^{max} = U_{te}^{max} CAP_{te} \qquad \forall te \tag{17}$$

$$(x_{te})_s (CAP_{te}^{max})^{-1} = (B_{te})_s \qquad \forall s, \forall te \tag{18}$$

In this work, the milling process is optimised with the objective function given in Equation (1) by deactivating the material input constraint, Equation (8) to determine the maximum product output of the system, y_m^{max}. At this point, the technology bottleneck of the system is indicated by B_{te} equal to one ($B_{te} = 1$), representing that a particular technology has been fully utilised, capping the y_m of the entire system. It is assumed that process intensification of the technology bottleneck is not possible and additional equipment unit will be needed to increase y_m^{max}, where B_{te} serves as an indicator to pinpoint the additional technology equipment for purchase/upgrade.

Following that, the objective function is modified into Equation (19) to determine the minimum output of the system, y_m^{min} while ensuring the system is economically stable to sustain its operation (i.e., *EP* equal to zero). In the event where minimum *EP* is required at a targeted value, additional constraints may be added to the formulation. The changes in y_m^{max} and y_m^{min} are measured for each incremental step in summation of U_{te}^{max} ($\sum_{te=1}^{TE} U_{te}^{max}$) by one equipment unit at a time to determine the feasible operating range of each design.

$$\text{Minimise } EP \tag{19}$$

The utilisation index, *UI* and flexibility index, *FI* for each incremental unit of $\mathbf{U}_{te}^{max}$ is determined via Equations (20) and (21) to measure the operational performance of the system

$$(UI)_s = \frac{(y_m)_s}{y_m^{max}} \qquad \forall s \tag{20}$$

$$(FI)_s = \frac{y_m^{max} - (y_m)_s}{y_m^{max}} \qquad \forall s \tag{21}$$

where *UI* and *FI* range between zero to one. In the event where *UI* equals to zero, the process is not utilised while *UI* equals to one indicates that the process is operating at 100% of the processing capacity installed. Meanwhile, zero in *FI* represents that the process has no flexibility in its operation and vice versa. To better illustrate the proposed FORA, a generic process where y_m^{max}, y_m^{min}, *UI* and *FI* are plotted against *CAPEX* as shown in Figure 2a. Several key features to be highlighted from the analysis are as follows:

1. Cross and plus markers in Figure 2a represent y_m^{max} and y_m^{min} of different system design with different $\mathbf{U}_{te}^{max}$ (x-axis on the left) and a corresponding *CAPEX* required (y-axis). The area shaded in grey between y_m^{max} and y_m^{min} represented the feasible operating range of the developed system where y_m (yellow line) must fall in between this region. This is to ensure that the system output is always less than or equal to the maximum production capacity, while greater or equal to the minimum output to sustain its operation.

2. y_m^{max} and y_m^{min} changes with the system design, during the addition or removal of the equipment unit. Hence, step changes are observed in y_m^{max} and y_m^{min} (black lines) when *CAPEX* increases.

3. The technology bottleneck and additional equipment unit to be added for each step are identified by B_{te} when $B_{te} = 1$. Increments in y_m^{max} and y_m^{min} are not proportional to the increment in *CAPEX* as the capital and capacity of each technology varies according to the market. Occasionally, more than one technology bottlenecks might occur. In that respect, multiple types of equipment and greater *CAPEX* are needed to increase the capacity of the system.

4. The increment in y_m^{max} (black line) reduces the *UI* (green line) while increases the *FI* (blue line) of the system in the same behaviour due to the changes in production capacity as shown in the x-axis on the right.

In the event where y_m is increased to y_{te}^{max}, the first design becomes infeasible (area shaded in red) as y_m^{new} falls out of the area shaded in grey as shown in Figure 2b. At the same time, a budget constraint is applied where only an increment up to a maximum capital cost, CAPEX^{max} can be invested. Based on the diagram, design 2 is required to cope up with such changes and *CAPEX*$_2$ falls within the constraint (*CAPEX*$_2$ < CAPEX^{max}). As such, the expansion from first to second design is feasible and the system will have more flexibility in product output up to y_{m2}^{max} level. On the other hand, when y_m^{new} is further increased as shown in Figure 2c, the *CAPEX* required falls beyond the constraint (*CAPEX*$_3$ > CAPEX^{max}). This shows that none of the design is suitable for such increment in y_m. Hence, the decision maker may only expand up to y_{m2}^{max} with *CAPEX*$_2$, sacrificing the additional output between y_m^{new} and y_{m2}^{max}.

Besides targeting for changes in y_m, the proposed approach allows for decision making based on *FI* or *UI*. For instance, the same investor is interested in changing the system slightly, allowing for more flexibility, FI^{new} to tolerate greater fluctuation in y_m which may occur in the future. Similarly, budget constraint is capped at $CAPEX^{max}$. As presented in Figure 2d, FI^{new} is greater than FI_1 but lesser FI_2. Therefore, flexibility up to FI_2 can be achieved based on the constraint set. This analysis serves as a powerful tool to plan for increment in product output expected in the future, under different constraints.

The following section presents a typical POM case study in Malaysia to illustrate the proposed approach. An IOM is developed to optimise the milling process, followed by FORA to study the feasible operating range within its design capacity. The developed mixed-integer linear programming model was solved via LINGO (v16, LINDO Systems, Inc., Chicago, USA) to achieve a global solution [45], with an Intel® Core™ i5 (2 × 3.20 GHz); 8 GB DDR3 RAM desktop unit. Alternatively, other optimisation software such as MATLAB and Statistics Toolbox (Release 2012b, The MathWorks, Inc., Natick, MA, USA) and General Algebraic Modeling System (GAMS) (Release 24.2.1, GAMS Development Corporation, Washington, DC, USA) could be used to achieve the same solution, depending on user preference.

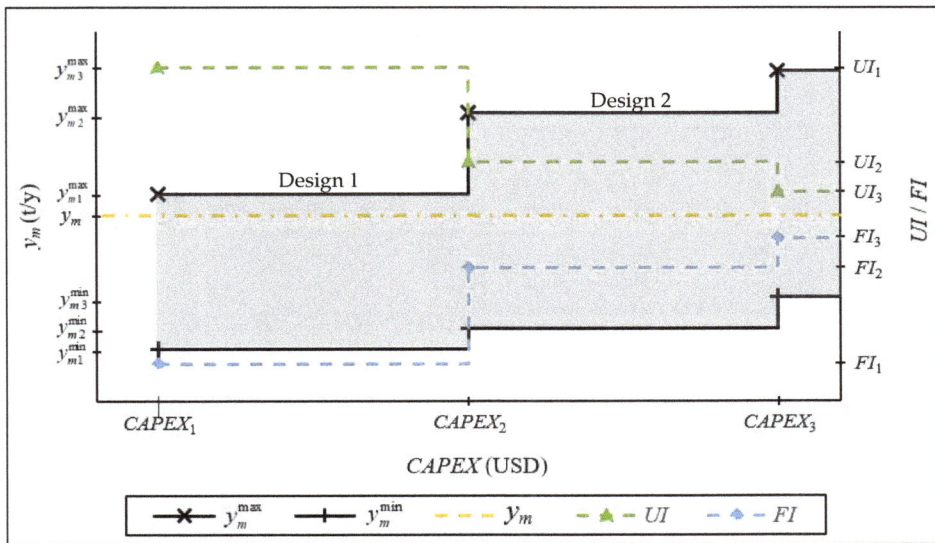

(a)

Figure 2. *Cont.*

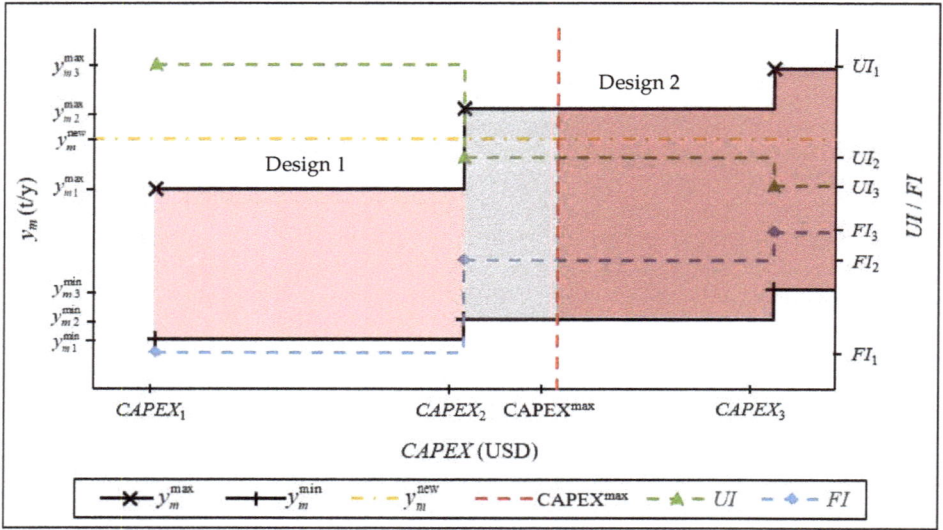

(b)

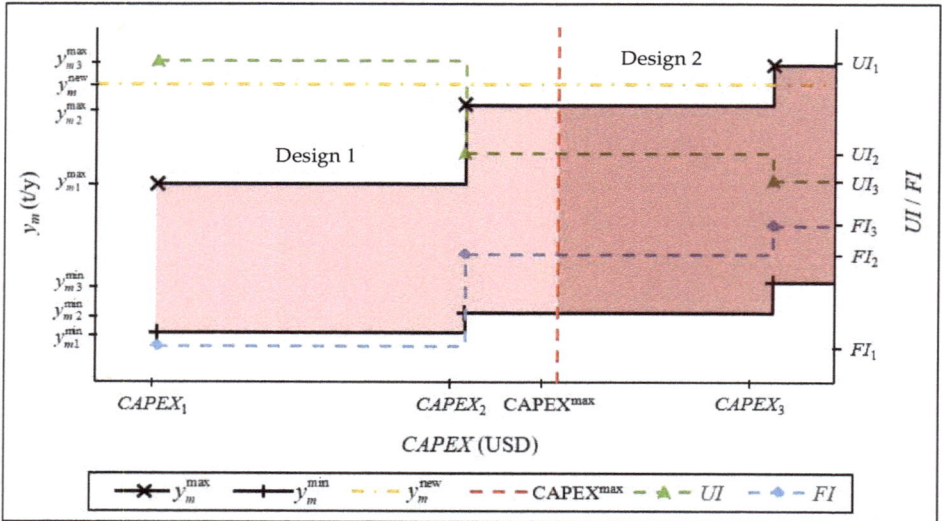

(c)

Figure 2. *Cont.*

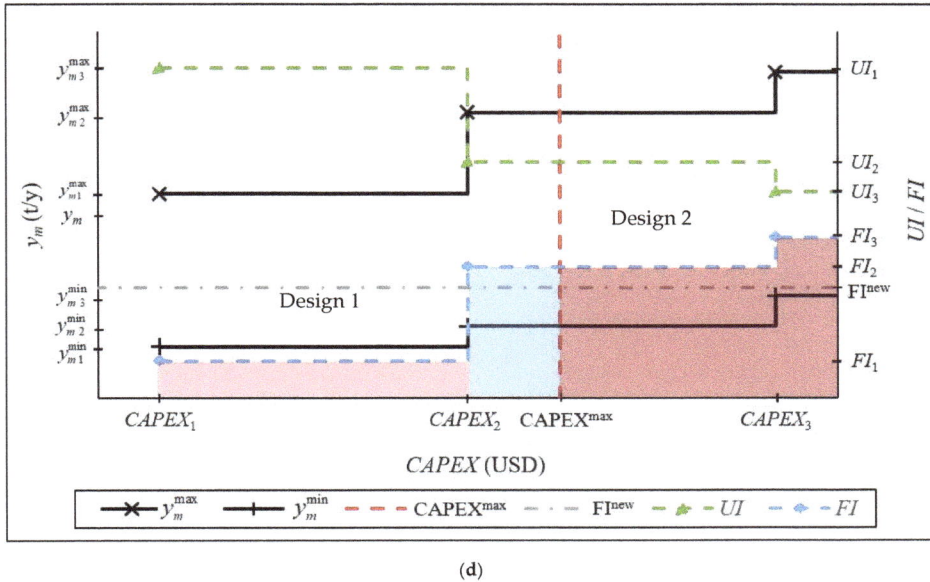

(**d**)

Figure 2. (**a**) FORA for a generic process. (**b**) FORA with y_m^{new} within the CAPEX^{max} constraints. (**c**) FORA with y_m^{new} exceeding the CAPEX^{max} constraints. (**d**) FORA with FI^{new} and CAPEX^{max} constraints.

4. Case Study

In this case study, the hybrid approach presented is demonstrated using a POM design adopted from Foong et al. [6] as the baseline design. It is assumed that the mill operator is interested to further optimise the milling process to improve economic performance, *EP* by taking operational factors such as operating hours, labour costs and FFB availability into account. Besides, an analysis to study the feasible operating range, utilisation and flexibility of the POM design is performed, providing a better insight for any changes in system design to cater for any variation in production output in the future. FFBs obtained from plantations are divided into three crop seasons, that is, low, medium and high seasons, each with a given fraction of occurrence, α_s and availability, y_{FFB} as shown in Table 1.

Table 1. The fraction of occurrence and FFB availability for different crop seasons.

Crop Season	Fraction of Occurrence, α_s	FFB Availability, y_{FFB} (t/y)
Low	$\alpha_L = 0.417$	$-195,800$
Medium	$\alpha_M = 0.333$	$-261,000$
High	$\alpha_H = 0.250$	$-369,800$
Average		$-261,000$

Reprinted (adapted) with permission from [6], copyright (2018) American Chemical Society.

A typical POM operates in batches for 12 h daily, usually divided into two workings shifts (i.e., annual shift time, AST = 4350 h/y). It is assumed that the POM is located in a remote area where power grid connection is not available and electricity required to operate the milling process is produced by cogeneration of biomass resources such as PPF and PKS. Fifteen operators with a labour cost, C$_{lab}$ of US\$4500/y is required for each shift to operate the milling process. It is further assumed that the POM will have an operation lifespan, t_{te}^{max} of 15 years with a discount rate, r of 5% per annum. The baseline POM design is shown in Figure 3 with the material and energy flows reported in a range for low and high crop seasons while the values stated in bracket represents the equipment units of each technology needed. Economic parameters such as *CAPEX, OPEX, LC, GP* and *EP* are summarised in Table 2 while additional information on material flows, technology units, process matrix table and other specifications of the system (i.e., $\mathbf{CAP}_{te}$, $\mathbf{E}_{te}$, $\mathbf{CC}_{te}$, $\mathbf{OC}_{te}$ and $\mathbf{C}_m$) provided in Tables 3–7.

Table 2. Economic parameters for baseline POM design.

Economic Parameters	Low Season	Medium Season	High Season	Average
Total capital costs, *CAPEX* (million US$)		18.42		18.42
Annualised *CAPEX* (million US$/y)		1.77		1.77
Labour costs, *LC* (million US$/y)		0.14		0.14
Total operating costs, *OPEX* (million US$/y)	1.13	1.33	1.87	1.38
Gross Profit, *GP* (million US$/y)	3.89	5.64	8.10	5.53
Economic Performance, *EP* (million US$/y)	2.12	3.87	6.32	3.75

Reprinted (adapted) with permission from [6], copyright (2018) American Chemical Society.

Table 3. Material and energy flows for baseline POM design.

Material Flows	Low Season		Medium Season		High Season		Average	
	(t/h)	(t/y)	(t/h)	(t/y)	(t/h)	(t/y)	(t/h)	(t/y)
Low pressure steam, LPS	−15.3	−66,600	−20.3	−88,300	−28.8	−125,300	−20.3	−88,500
Utility water	−13.4	−58,300	−17.8	−77,400	−25.3	−110,000	−17.8	−77,500
Crude palm oil, CPO	9.3	40,500	12.4	54,000	17.6	76,600	12.4	54,000
Palm kernel, PK	3.3	14,500	4.5	19,500	6.4	28,000	4.5	19,500
Palm pressed fibre, PPF	0	0	0	0	0	0	0	0
Palm kernel shell, PKS	1.4	6000	2.9	12,800	4.3	18,800	2.6	11,500
Pressed empty fruit bunch, PEFB	8.5	37,000	11.3	49,000	15.9	69,000	11.3	49,000
Palm oil mill effluent, POME	31.3	136,000	41.7	181,500	59.1	257,000	41.7	181,500
Energy Flows	**(kW)**	**(MWh/y)**	**(kW)**	**(MWh/y)**	**(kW)**	**(MWh/y)**	**(kW)**	**(MWh/y)**
Electricity demand, E^{Demand} (kW)	990	4292	1100	4967	1600	6933	1200	5178

Reprinted (adapted) with permission from [6], copyright (2018) American Chemical Society.

Table 4. Technology units operated for baseline POM design.

Technology Units Operated (units)	Low Season	Medium Season	High Season
Tilted steriliser	3	3	5
Rotating drum separator	1	2	2
Oil pressing screw	1	2	2
Steam injection digester	2	3	3
Double screw press	2	2	3
Depricarper	2	2	3
Rolek nut cracker	1	2	2
Four-stage winnowing column	1	1	1
Vertical clarifier	2	3	4
Vacuum dryer	2	2	3
Three-phase decanter	2	2	3
Oil recovery pit	1	2	2
Water tube boiler	1	2	2
High-pressure steam turbine	1	1	1
Medium-pressure steam turbine	2	2	3
Total	**24**	**31**	**39**

Reprinted (adapted) with permission from [6], copyright (2018) American Chemical Society.

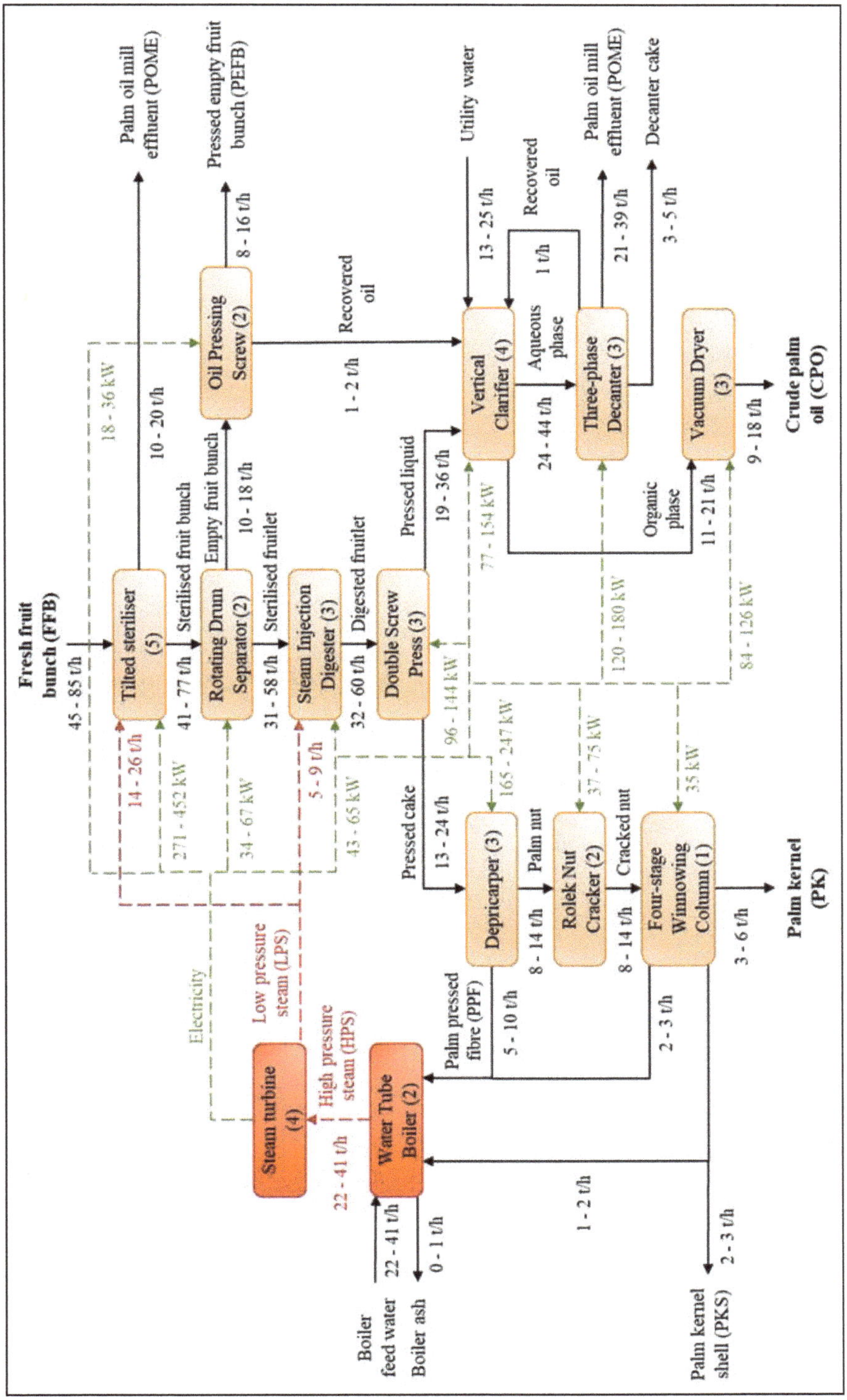

Figure 3. Baseline POM design.

Table 5. Process matrix A for palm oil milling process.

Material/Technology	te = 1 Tilted Steriliser (t/h)	te = 2 Rotating Drum Separator (t/h)	te = 3 Oil Pressing Screw (t/h)	te = 4 Steam Injection Digester (t/h)	te = 5 Double Screw Press (t/h)	te = 6 Depericarper (t/h)	te = 7 Rolek Nut Cracker (t/h)	te = 8 Four-Stage Winnowing Column (t/h)	te = 9 Vertical Clarifier (t/h)	te = 10 Oil Recovery (t/h)	te = 11 Vacuum Dryer (t/h)	te = 12 Three-Phase Decanter (t/h)	te = 13 Oil Recovery Pit (t/h)	te = 14 PPF Combustion (t/h)	te = 15 PKS Combustion (t/h)	te = 16 Water Tube Boiler (t/h)	te = 17 HPS Turbine (kW)	te = 18 MPS Turbine (kW)
m = 1 Fresh fruit bunch, FFB (t)	−1	0	0	0	0	0	0	0	0	0	0	0	0	0	0	0	0	0
m = 2 Utility water (t)	0	0	0	0	0	0	0	0	0	0	0	0	0	0	0	0	0	0
m = 3 Steam lost (t)	0.12	0	0	0.116	0	0	0	0	−0.696	0	0.138	0	0	0	0	0	0	0
m = 4 Sterilised fruit bunch (t)	0.9	−1	0	0	0	0	0	0	0	0	0	0	0	0	0	0	0	0
m = 5 Empty fruit bunch, EFB (t)	0	0.24	−1	0	0	0	0	0	0	0	0	0	0	0	0	0	0	0
m = 6 Sterilised fruitlet (t)	0	0.76	0	−1	0	0	0	0	0	0	0	0	0	0	0	0	0	0
m = 7 Digested fruitlet (t)	0	0	0	1.04	−1	0	0	0	0	0	0	0	0	0	0	0	0	0
m = 8 Pressed liquid (t)	0	0	0	0	0.6	0	0	0	−1	0	0	0	0	0	0	0	0	0
m = 9 Pressed cake (t)	0	0	0	0	0.4	−1	0	0	0	0	0	0	0	0	0	0	0	0
m = 10 Palm fruit nut (t)	0	0	0	0	0	0.59	−1	0	0	0	0	0	0	0	0	0	0	0
m = 11 Cracked nut (t)	0	0	0	0	0	0	0.99	−1	0	0	0	0	0	0	0	0	0	0
m = 12 Nut lost (t)	0	0	0	0	0	0	0.01	0	0	0	0	0	0	0	0	0	0	0
m = 13 Aqueous phase (t)	0	0	0	0	0	0	0	0	1.156	0	0	−1	0	0	0	0	0	0
m = 14 Organic phase (t)	0	0	0	0	0	0	0	0	0.54	0	−1	0	0	0	0	0	0	0
m = 15 Palm pressed fibre, PPF (t)	0	0	0	0	0	0.41	0	0.19	0	0	0	0	0	−1	0	0	0	0
m = 16 Palm kernel shell, PKS (t)	0	0	0	0	0	0	0	0.357	0	0	0	0	0	0	−1	0	0	0
m = 17 Pressed empty fruit bunch, PEFB (t)	0	0	0.868	0	0	0	0	0	0	0	0	0	0	0	0	0	0	0
m = 18 Decanter cake (t)	0	0	0	0	0	0	0	0	0	0	0	0	0	0	0	0	0	0
m = 19 Crude palm oil, CPO (t)	0	0	0	0	0	0	0	0	0	0.36	0.828	0.113	0	0	0	0	0	0
m = 20 Palm kernel, PK (t)	0	0	0	0	0	0	0	0.453	0	0	0	0	0	0	0	0	0	0
m = 21 Recovered oil (t)	0	0	0.132	0	0	0	0	0	0	−1	0	0.02	0.0097	0	0	0	0	0
m = 22 Palm oil mill effluent, POME (t)	0.23	0	0	0	0	0	0	0	0	0.64	0.034	0.867	−1	0	0	0	0	0
m = 23 Deoiled POME (t)	0	0	0	0	0	0	0	0	0	0	0	0	0.9903	0	0	0	0	0
m = 24 Boiler feed water (t)	0	0	0	0	0	0	0	0	0	0	0	0	0	0	0	−1	0	0
m = 25 Boiler ash (t)	0	0	0	0	0	0	0	0	0	0	0	0	0	0.0423	0.039	0	0	0
m = 26 Low heating value (MJ)	0	0	0	0	0	0	0	0	0	0	0	0	0	13388	17804	−5151.8	0	0
m = 27 Low pressure steam, LPS (t)	−0.3	0	0	−0.156	0	0	0	0	0	0	0	0	0	0	0	0	0	0.0316
m = 28 Medium pressure steam, MPS (t)	0	0	0	0	0	0	0	0	0	0	0	0	0	0	0	0	0.0735	−0.0316
m = 29 High pressure steam, HPS (t)	0	0	0	0	0	0	0	0	0	0	0	0	0	0	0	1	−0.0735	0
m = 30 Electricity (kW)	0	0	0	0	0	0	0	0	0	0	0	0	0	0	0	0	1	1

Table 6. Technology specifications for palm oil milling process.

Technology Specifications	$te=1$ Tilted Steriliser	$te=2$ Rotating Drum Separator	$te=3$ Oil Pressing Screw	$te=4$ Steam Injection Digester	$te=5$ Double Screw Press	$te=6$ Depricarper	$te=7$ Rolek Nut Cracker	$te=8$ Four-Stage Winnowing Column	$te=9$ Vertical Clarifier	$te=10$ Oil Recovery	$te=11$ Vacuum Dryer	$te=12$ Three-Phase Decanter	$te=13$ Oil Recovery Pit	$te=14$ PPF Combustion	$te=15$ PKS Combustion	$te=16$ Water Tube Boiler	$te=17$ HPS Turbine	$te=18$ MPS Turbine
Capacity, CAP_{te} (t/h.unit or kW/unit)	20	50	10	20	25	10	10	15	10	100	8	20	41	100	100	25	1000	500
Electricity, E_{te} (kW/unit)	75.4	28	15	18	40	69	31	29	32	0	35	50	5.5	0	0	0	0	0
Capital costs, CC_{te} (million US$/unit)	1.2	0.23	0.12	0.15	0.18	0.25	0.18	0.25	0.15	0	0.39	0.30	0.03	0	0	2.00	0.83	0.61
Operating costs, OC_{te} (million US$/unit.y)	0.18	0.03	0.02	0.02	0.04	0.03	0.04	0.01	0.02	0	0.06	0.04	0	0	0	0.08	0.02	0.01

Table 7. Material costs, C_m for palm oil milling process.

	$m = 1$ Fresh Fruit Bunch, FFB	$m = 2$ Utility Water	$m = 15$ Palm Pressed Fibre, PPF	$m = 16$ Palm Kernel Shell, PKS	$m = 16$ Palm Kernel Shell, PKS
Material Costs, C_m (US$/t)	121	0.55	23	45	45
	$m = 17$ Pressed Empty Fruit Bunch, PEFB	$m = 18$ Decanter Cake	$m = 19$ Crude Palm Oil, CPO	$m = 20$ Palm Kernel, PK	$m = 24$ Boiler Feed Water
	8	43	548	389	1.14

Intermediates associated with zero costs ($C_m = 0$) are not listed here.

The assumption that the milling process can only be operated for 4350 h a year due to the working shifts of operators causes its capacity to be underutilised. In that case, more equipment units are required, resulting in greater *CAPEX* needed to process all the FFBs supplied, especially during the peak crop season. This shows a limitation in the previous study [6] during optimisation of a palm oil milling process. A more common practice in the industry is to increase the annual operating time, *AOT* of the process. In the industry, POM may operate up to 19 h/day or 7000 h/y ($AOT \leq 7000$). In that sense, the total capital costs, *CAPEX* needed can be reduced as lesser equipment units are required. However, the increment in *AOT* on top of 4350 h/y AST requires overtime cost; *OTC* paid for operators working extra time and operating costs, OC_{te} for service and maintenance of technology units. In this study, overtime costs, C_{OT} of US$5/h and an additional 20% for OC_{te} are considered for operations exceeding 4350 h/y.

5. Results and Discussion

In order to achieve higher *EP*, an IOM was developed based on Equations (2)–(15) to optimise the baseline POM design with an objective function given in Equation (1). The model consists of 419 continuous variables with 54 integer variables and 622 constraints, solved in 17 s to achieve a global solution. The optimised POM design is presented in Figure 4 and the results (Table 8) showed that an *EP* of US$4.57 million/y *EP* is achieved (22% increment) as compared to US$3.75 million/y reported in the baseline design. This is mainly due to the reduction in *CAPEX* required, from US$18.42 to 11.56 million as the units of technology required, U_{te}^{max} reduce from 39 (Table 4) to 26 units as shown in Table 9.

Data from Table 10 is compared with Table 3, showing that the same annual output is achieved, despite a smaller throughput (material flow per hour) in the optimised design by operating 5580, 5640, 4698 and 6656 h/y on average, low, medium and high crop seasons, respectively. In this respect, additional *OTC* by US$0.10, 0.03 and 0.17 million/y required for different crop seasons (an average of US$0.09 million/y). Besides, an additional 20% OC_{te} is required to operate the equipment due to longer operational time. However, *OPEX* is still reduced by US$0.32 million/y on average (= US$1.38–1.06 million/y) as the overall equipment operated decreases. It is worth mentioning that the equipment operated is the same for medium and high crop seasons but longer *AOT* in the latter case. As higher *OTC* is required to process all the fruits available during medium crop season with smaller processing capacity, it is more optimal to operate the milling process with higher throughput but lower *AOT*. Figure 5 shows the breakdown of costs allocation for both designs where *CAPEX* is annualised into a yearly basis for 15 years. It can be seen that the total costs required by the optimised design are lower than the baseline design by to 23% on average with 25, 18 and 25% reduction during low, medium and high crop seasons, respectively. In the next part of this section, the milling process is further analysed with FORA as mentioned earlier to study the feasible range of CPO output with respect to *CAPEX* invested.

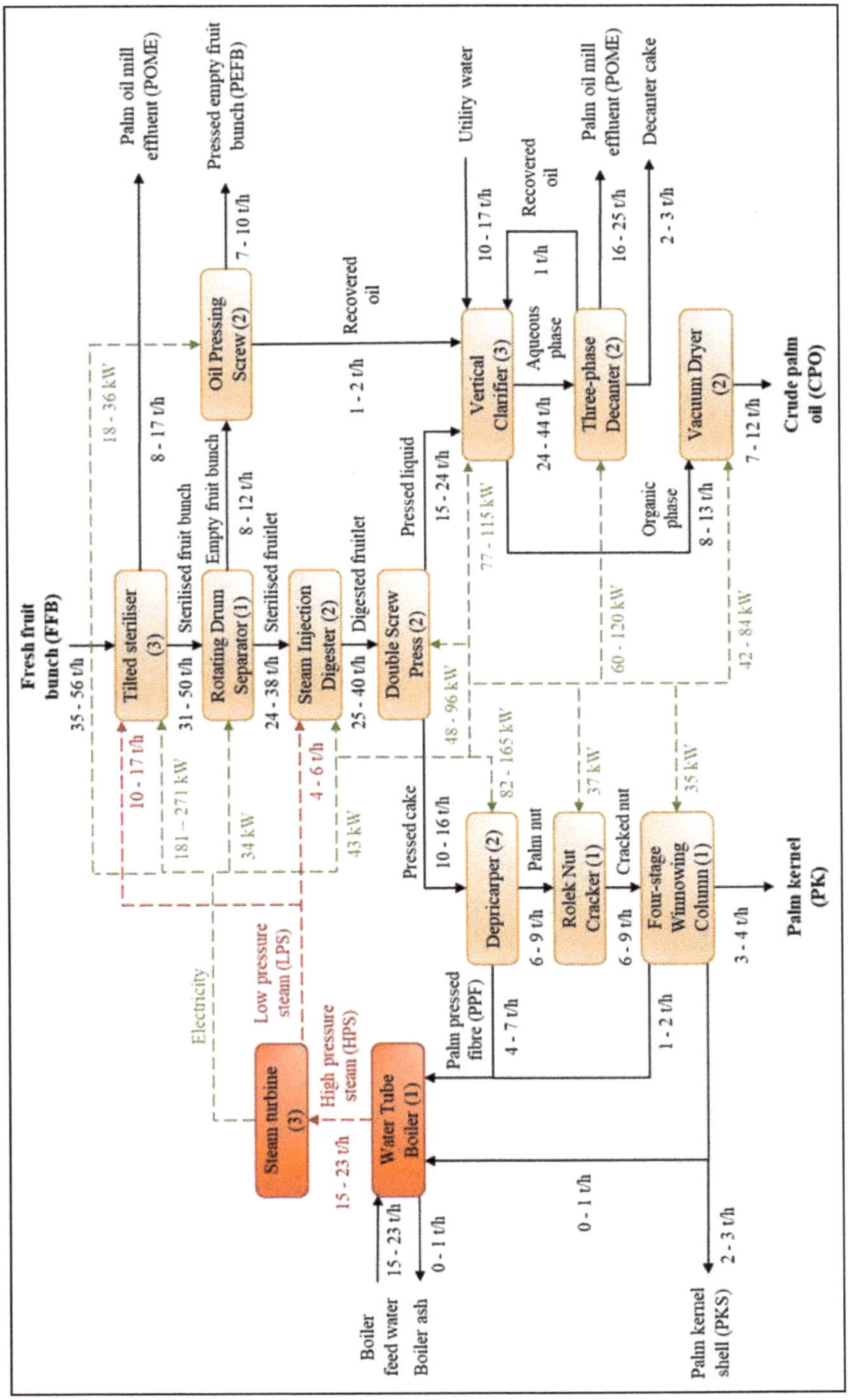

Figure 4. Optimised POM design.

Table 8. Economic parameters for optimised POM design.

Economic Parameters	Low Season	Medium Season	High Season	Average
Annual operational time, AOT (h/y)	5640	4700	6660	6520
Total capital costs, $CAPEX$ (million US$)		11.56		11.56
Annualised $CAPEX$ (million US$/y)		1.11		1.11
Labour costs, LC (million US$/y)		0.14		0.14
Overtime costs, OTC (million US$/y)	0.10	0.03	0.17	0.09
Total operating costs, $OPEX$ (million US$/y)	0.94	1.40	1.40	1.21
Gross Profit, GP (million US$/y)	4.15	5.57	8.39	5.68
Economic Performance, EP (million US$/y)	3.04	4.46	7.28	4.57

Table 9. Technology units operated for optimised POM design.

Technology Units Operated (Units)	Low Season	Medium Season	High Season
Tilted steriliser	2	3	3
Rotating drum separator	1	1	1
Oil pressing screw	1	2	2
Steam injection digester	2	2	2
Double screw press	1	2	2
Depricarper	1	2	2
Rolek nut cracker	1	1	1
Four-stage winnowing column	1	1	1
Vertical clarifier	2	3	3
Vacuum dryer	1	2	2
Three-phase decanter	1	2	2
Oil recovery pit	1	1	1
Water tube boiler	1	1	1
High-pressure steam turbine	1	1	1
Medium-pressure steam turbine	1	2	2
Total	**18**	**26**	**26**

Table 10. Material and energy flows for optimised POM design.

Material Flows	Low Season		Medium Season		High Season		Average	
	(t/h)	(t/y)	(t/h)	(t/y)	(t/h)	(t/y)	(t/h)	(t/y)
Low pressure steam, LPS	−11.8	−66,600	−18.8	−88,300	−18.8	−125,300	−15.9	−88,500
Utility water	−10.3	−58,300	−16.5	−77,400	−16.5	−110,000	−13.9	−77,500
Crude palm oil, CPO	7.2	40,500	11.5	54,000	11.5	76,600	9.7	54,000
Palm kernel, PK	2.6	14,500	4.2	19,500	4.2	28,000	3.5	19,500
Palm pressed fibre, PPF	0	0	0	0	0	0	0	0
Palm kernel shell, PKS	1.7	9500	1.7	7900	2.8	18,800	2.3	13,000
Pressed empty fruit bunch, PEFB	8.5	37,000	10.4	49,000	10.4	69,000	8.8	49,000
Palm oil mill effluent, POME	24.1	136,000	38.6	181,500	38.6	257,000	32.5	181,500
Energy Flows	(kW)	(MWh/y)	(kW)	(MWh/y)	(kW)	(MWh/y)	(kW)	(MWh/y)
Electricity demand, E^{Demand} (kW)	660	3744	1000	4900	1000	6940	890	4938

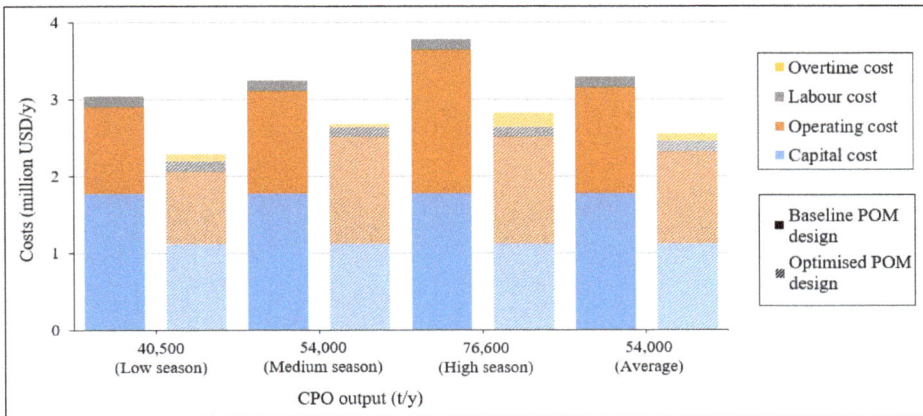

Figure 5. Costs allocation for a baseline against optimised palm oil mill design.

FORA is performed on the milling process based on Equations (16)–(18), subject to objective functions Equations (1) and (19) for each POM design while operational performance such as *UI* and *FI* are computed based on Equations (20)–(21). The analysis is performed for each increment in equipment unit added, beginning from the design with the smallest capacity of 18 units (optimised design during low crop season) to the design with the biggest capacity, 39 units in the baseline design. Graphical representations for FORA performed on different POM designs are presented in Figure 6 for different crop seasons and detailed information can be found in Table 11. From Figure 6a, we can see that the CPO production during low crop season, $(y_{CPO})_L$ falls within the entire feasible region, representing that each of the design can be used to achieve the output required. In this respect, the optimal operation will be determined from the trade-off between *OTC*, *CAPEX* and *OPEX* as a design with smaller capacity requires higher *OTC* but lower *CAPEX* and *OPEX* or vice versa. According to Tables 8–10, the POM is operated at smallest design capacity (*CAPEX* = US$8.36 million) with longer *AOT* of 5640 h/y (*OTC* = US$0.1 million/y) during the low crop season. However, the POM is operated in a different manner during medium crop season. Figure 6b shows that $(y_{CPO})_M$ lies in the feasible range for POM designs with 21 equipment units (*CAPEX* = US$9.18 million) and higher. Rather than operating the process with the smallest capacity possible, it was operated at a higher capacity (26 equipment units, *CAPEX* = US$11.56 million) due to lower *OTC* of US$0.03 million/y (*AOT* = 4700). On the other hand, at least 26 equipment units are needed during high crop season as $(y_{CPO})_H$ falls out of the feasible operating range for smaller POM design as shown in Figure 6c. From the optimised results, the smallest possible design with higher *OTC* of US$0.17 million/y (*AOT* = 6660) is operated during this crop season.

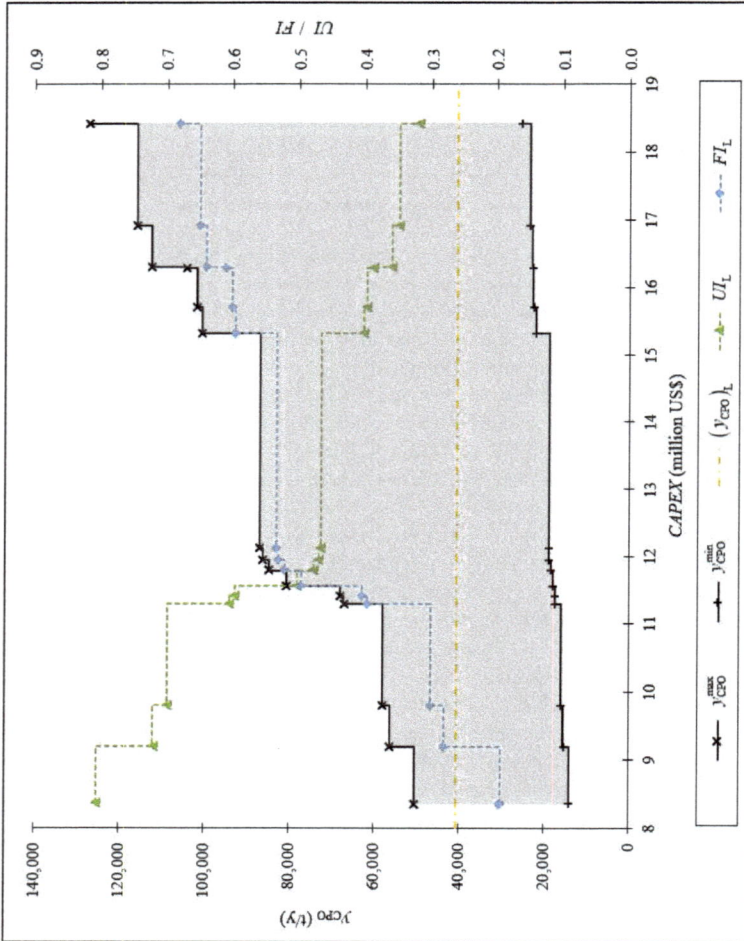

(a)

Figure 6. *Cont.*

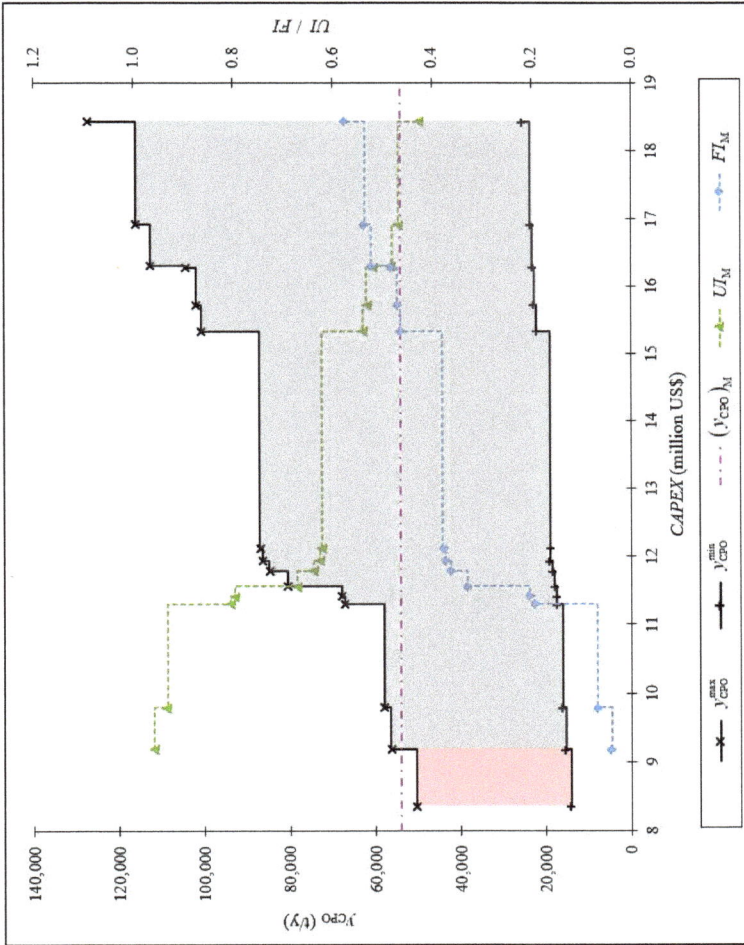

(b)

Figure 6. *Cont.*

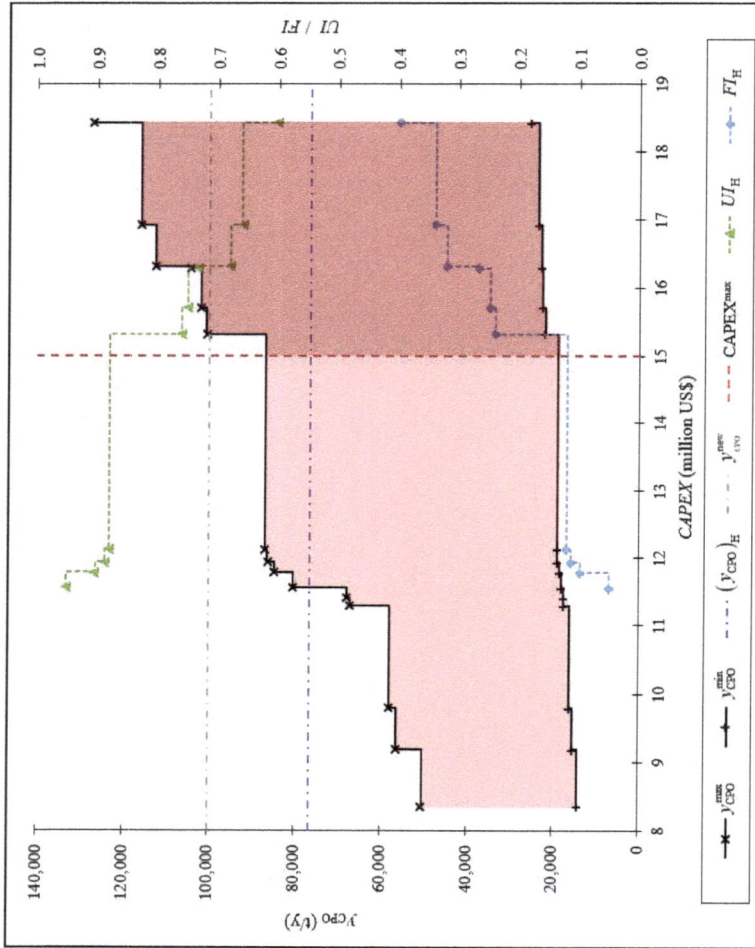

(c)

Figure 6. (a) Graphical representations for FORA performed during low crop season. (b) Graphical representations for FORA performed during medium crop season. (c) Graphical representations for FORA performed during high crop season with y_{CPO}^{new}.

Table 11. Feasible operating range analysis data.

u_{te} (units)	y_{CPO}^{max} (t/y)	y_{CPO}^{min} (t/y)	CAPEX (million US$)	Additional CAPEX (million US$)	Equipment Added	CBR	UI_L	FI_L	UI_M	FI_M	UI_H	FI_H
18	50,300	14,100	8.36	-	Vacuum dryer	-	0.81	0.19	-	-	-	-
21	56,300	15,400	9.18	0.82	Double screw press Depricarper	3.5	0.72	0.28	0.96	0.04	-	-
22	58,000	16,100	9.79	0.61	MPS turbine	2.9	0.70	0.30	0.93	0.07	-	-
24	67,000	17,500	11.29	1.50	Three-phase decanter Tilted steriliser	3.0	0.60	0.40	0.81	0.19	-	-
25	68,000	17,500	11.41	0.12	Oil pressing screw	3.1	0.60	0.40	0.80	0.20	-	-
26	80,500	18,000	11.56	0.15	Steam injection digester	44.3	0.50	0.50	0.67	0.33	0.95	0.05
27	84,800	18,500	11.79	0.23	Rotating drum separator	9.9	0.48	0.52	0.64	0.36	0.90	0.10
28	86,300	19,000	11.94	0.15	Vertical clarifier	5.5	0.47	0.53	0.63	0.37	0.89	0.11
29	87,000	19,000	12.12	0.18	Rolek nut cracker	1.3	0.47	0.53	0.62	0.38	0.88	0.12
31	100,600	22,000	15.32	3.21	Tilted steriliser	2.5	0.40	0.60	0.54	0.46	0.76	0.24
32	101,900	22,700	15.49	0.18	Water tube boiler Double screw press Depricarper	2.7	0.40	0.60	0.53	0.47	0.75	0.25
35	104,400	22,900	16.29	0.80	Vacuum dryer Vertical clarifier	1.6	0.39	0.61	0.52	0.48	0.73	0.27
36	112,600	23,000	16.31	0.02	Oil recovery pit	415.3	0.36	0.64	0.48	0.52	0.68	0.32
37	116,000	23,600	16.92	0.61	MPS turbine	5.9	0.35	0.65	0.47	0.53	0.66	0.34
39	127,200	25,500	18.42	1.50	Three-phase decanter Tilted steriliser	3.7	0.32	0.68	0.43	0.57	0.60	0.40

Apart from determining the feasible operating range of each design, this approach also serves as a tool to pinpoint the technology bottleneck, additional *CAPEX* needed and y_{CPO}^{max} increment for the milling process in sequence. Table 11 shows that additional equipment units for vacuum dryer, double screw press and depricarper technologies are needed to increase y_{CPO}^{max} from 50,300 to 56,300 t/y. Three different bottlenecks occur at the same time and y_{CPO}^{max} can only be increased when all three equipment units added. It is then followed by MPS turbine and three-phase decanter to increase y_{CPO}^{max} from 56,300 to 58,000 t/y and so on. It also allows the cost-benefit ratio, *CBR* for each step to be performed via Equation (22), providing more insight into the effectiveness of any additional investment made.

$$CBR = \frac{C_{CPO}\left(y_{CPO}^{max\,2} - y_{CPO}^{max\,1}\right)}{CRF(CAPEX_2 - CAPEX_1) + (OPEX_2 - OPEX_1)} \tag{22}$$

Based on the y_{CPO} for each crop season, the *UI* and *FI* vary with its design. Note that *UI* and *FI* can only be measured when y_{CPO} falls within the feasible operating range. During high crop season, a more significant portion of the production capacity in the optimised design has been utilised (*UI* = 0.95) as compared to the baseline design with *UI* of 0.60. However, it reduces the flexibility from *FI* of 0.40 to 0.05. This indicates that even though a higher proportion of the production capacity utilised in the optimised design during high crop season, the flexibility is reduced. In the event where y_{CPO} were to increase further, it is implausible for the optimised design to cope up with such changes, unless, additional equipment units for rotating drum separator, vertical clarifier, rolek nut cracker and so forth are added. For instance, when $(y_{CPO})_H$ is increased by 30% from 76,600 to 100,000 t/y, *CAPEX* of US\$15.32 million and 31 equipment units will be needed to achieve the y_{CPO}^{new} as shown in Figure 6c. However, such increment could not be satisfied if $CAPEX^{max}$ is limited at US\$15 million. Thus, a maximum of 87,000 t/y CPO could be produced with such given constraint in *CAPEX*.

6. Conclusions

A hybrid methodology was developed in this work to optimise a typical palm oil milling process to achieve maximum economic performance, performing an analysis for its operations and providing a feasibility study on the developed system. This hybrid approach consists of generic formulations for IOM and FORA to represent a palm oil milling process. The proposed approach has been illustrated using a Malaysian palm oil mill case study with multiple crop seasons. In the case study, higher *EP* is achieved from the optimised POM design with a smaller capacity but longer operational time as compared to the baseline design used. The utilisation of the POM has been improved. However, the flexibility of the process is also reduced proportionally. FORA serve as a decision-making tool to determine the *CAPEX* required, based on the output required with other constraints considered. Future research work will be directed to consider partial load models for changes in power consumption and process efficiency of each equipment units, analysing the detailed performance of each equipment and possibility for process intensification. Besides, sensitivity analysis for selling electricity, uncertainties in product prices and raw material availability due to external reasons and FORA formulation with multiple products output as well as the integration of downstream processes such as biorefinery can be included for decision-making in the future.

Author Contributions: Conceptualization, S.Z.Y.F. and D.K.S.N.; methodology, S.Z.Y.F.; software, S.Z.Y.F. and D.K.S.N.; validation, V.A., R.R.T. and D.C.Y.F.; formal analysis, S.Z.Y.F., V.A. and D.K.S.N.; investigation, R.R.T. and D.C.Y.F.; resources, D.K.S.N.; data curation, S.Z.Y.F.; writing—original draft preparation, S.Z.Y.F.; writing—review and editing, V.A., R.R.T. and D.C.Y.F.; visualisation, S.Z.Y.F.; supervision, D.K.S.N.; funding acquisition, D.K.S.N.

Funding: This research was funded by the Ministry of Higher Education, Malaysia through LRGS Grant (LRGS/2013/UKM-UNMC/PT/05).

Acknowledgments: Credit to Havys Oil Mill Sdn Bhd for the industrial data provided to develop an industrial case study in this work.

Conflicts of Interest: The authors declare no conflict of interest. The funders had no role in the design, analyses and interpretation of any data of the study.

Nomenclature

Abbreviation

CPKO	Crude Palm Kernel Oil
CPO	Crude Palm Oil
FFB	Fresh Fruit Bunch
FORA	Feasible Operating Range Analysis
IO	Input-output
IOM	Input-output Optimisation Model
PEFB	Pressed Empty Fruit Bunch
PK	Palm Kernel
PKS	Palm Kernel Shell
POM	Palm Oil Mill
POME	Palm Oil Mill Effluent
PPF	Palm Pressed Fibre

Sets

H	Index for high crop season
L	Index for low crop season
M	Index for medium crop season
m	Index for material
s	Index for crop season
te	Index for technology

Variables

AOT	Annual operational time
B_{te}	Technology te bottleneck
$CAPEX$	Total capital costs
$CAPEX_1$	Total capital costs for design 1
$CAPEX_2$	Total capital costs for design 2
$CAPEX_3$	Total capital costs for design 3
CBR	Cost-benefit ratio
CRF	Capital recovery factor
E^{Demand}	Total electricity demand
EP	Economic performance
FI	Flexibility index
GP	Total gross profit
LC	Total labour costs
$OPEX$	Total operating costs
OTC	Total overtime costs
UI	Utilisation index
U_{te}	Number of equipment unit operated for technology te
x_{te}	Processing capacity of technology te

x_{te}	Processing capacity of technology *te*
y_{CPO}	Crude palm oil output
y_{CPO}^{new}	New crude palm oil output
$y_{electricity}$	Electricity output
y_{FFB}	Fresh fruit bunch input
y_m	Input or output of material *m*
y_{CPO}^{max}	Maximum crude palm oil output
y_m^{max}	Maximum input or output of material *m*
y_{m1}^{max}	Maximum input or output of material *m* for design 1
y_{m2}^{max}	Maximum input or output of material *m* for design 2
y_{m3}^{max}	Maximum input or output of material *m* for design 3
y_{CPO}^{min}	Minimum crude palm oil output
y_m^{min}	Minimum input or output of material *m*
y_{m1}^{min}	Minimum input or output of material *m* for design 1
y_{m2}^{min}	Minimum input or output of material *m* for design 2
y_{m3}^{min}	Minimum input or output of material *m* for design 3
y_m^{new}	New input or output of material *m*

Parameters

α_s	Fraction of occurrence for crop season *s*
A	Matrix of material input and output ratios to and from technology *te*
$a_{m,te}$	Fixed interaction ratios between material *m* and technology *te*
AOT^{max}	Maximum annual operating time
AST	Annual shift time
$CAPEX^{max}$	Maximum total capital costs
$\mathbf{CAP}_{te}$	Nominal capacity of technology *te*
$\mathbf{CAP}_{te}^{max}$	Maximum capacity of technology *te*
$(\mathbf{CAP}_{te}^{max})^{-1}$	Inverse matrix for maximum capacity of technology *te*
$\mathbf{CAP}_{te}^{-1}$	Inverse matrix for nominal capacity of technology *te*
$\mathbf{CC}_{te}$	Capital cost for technology *te*
C_{lab}	Cost of material *m*
C_m	Total overtime costs
C_{OT}	Specific overtime cost
$\mathbf{E}_{te}$	Diagonal matrix for electricity consumption specified per unit technology *te*
FI^{new}	New flexibility index
n_{wk}	Number of workers per shift
n_{ws}	Number of working shifts per day
$\mathbf{OC}_{te}$	Operating and maintenance costs for technology *te*
r	Discount rate
t_{te}^{max}	Operational lifespan for technology *te*
$\mathbf{U}_{te}^{max}$	Maximum units of technology *te* installed

References

1. Petrenko, C.; Paltseva, J.; Searle, S. *Ecological Impacts of Palm Oil Expansion in Indonesia*; International Council on Clean Transportation: Washington, DC, USA, 2016.
2. Reeves, J.B.; Weihrauch, J.L. *Composition of Foods: Fats and Oils: Raw, Processed, Prepared*; Agricultur.; United States Department of Agriculture Science and Education Administration: Washington, DC, USA, 1979.
3. Varkkey, H.; Tyson, A.; Choiruzzad, S.A.B. Palm oil intensification and expansion in Indonesia and Malaysia: Environmental and socio-political factors influencing policy. *For. Policy Econ.* **2018**, *92*, 148–159. [CrossRef]
4. Jekayinfa, S.O.; Bamgboye, A.I. Energy Requirements for Palm-kernel Oil Processing Operations. *Nutr. Food Sci.* **2004**, *34*, 166–173. [CrossRef]
5. Ng, R.T.L.; Ng, D.K.S. Systematic Approach for Synthesis of Integrated Palm Oil Processing Complex. Part 1: Single Owner. *Ind. Eng. Chem. Res.* **2013**, *52*, 10206–10220. [CrossRef]
6. Foong, S.Z.Y.; Lam, Y.L.; Andiappan, V.; Foo, D.C.Y.; Ng, D.K.S. A Systematic Approach for the Synthesis and Optimization of Palm Oil Milling Processes. *Ind. Eng. Chem. Res.* **2018**, *57*, 2945–2955. [CrossRef]
7. Umar, M.S.; Jennings, P.; Urmee, T. Sustainable electricity generation from oil palm biomass wastes in Malaysia: An industry survey. *Energy* **2014**, *67*, 496–505. [CrossRef]
8. Abdullah, N.; Sulaiman, F. The Oil Palm Wastes in Malaysia. In *Biomass Now: Sustainable Growth and Use*; Matovic, M.D., Ed.; InTech: Rijeka, Croatia, 2013; pp. 75–93.
9. Yusniati; Parinduri, L.; Sulaiman, O.K. Biomass Analysis at Palm Oil Factory as an Electric Power Plant. *J. Phys. Conf. Ser.* **2018**, *1007*, 12053. [CrossRef]
10. Umar, M.S.; Urmee, T.; Jennings, P. A policy framework and industry roadmap model for sustainable oil palm biomass electricity generation in Malaysia. *Renew. Energy* **2018**, *128*, 275–284. [CrossRef]
11. Husain, Z.; Zainal, Z.A.; Abdullah, M.Z. Analysis of Biomass-Residue-Based Cogeneration System in Palm Oil Mills. *Biomass Bioenergy* **2003**, *24*, 117–124. [CrossRef]
12. Zafar, S. Energy Potential of Palm Kernel Shells. Available online: https://www.bioenergyconsult.com/tag/pks-market-trends/ (accessed on 4 August 2018).
13. Husain, Z.; Zainac, Z.; Abdullah, Z. Briquetting of Palm Fibre and Shell from the Processing of Palm Nuts to Palm Oil. *Biomass Bioenergy* **2002**, *22*, 505–509. [CrossRef]
14. van Dam, J.E.G.; Elbersen, H.W. Palm Oil Production for Oil and Biomass: The Solution for Sustainable Oil Production and Certifiably Sustainable Biomass Production? 2004. Available online: http://library.wur.nl/WebQuery/wurpubs/350255 (accessed on 4 August 2018).
15. Yusuf, E.A. Development and Evaluation of Organic Biofertilizer Using Oil Palm Empty Fruit Bunch as Composted Medium for Vegetable Gardening. Ph.D. Thesis, University Malaysia Sarawak, Kota Samarahan, Malaysia, 2014.
16. Tong, S.L.; Jaafar, A.B. Waste to Energy: Methane Recovery from Anaerobic Digestion of Palm Oil Mill Effluent. *Energy Smart* **2004**, *4*, 1–8.
17. Yu-Lee, R.T. *Essentials of Capacity Management*; Wiley: Hoboken, NJ, USA, 2002; ISBN 0471207462.
18. Mohamad, F.; Mat Tahar, R.; Awang, N. Evaluating Capacity Of Palm Oil Mill Using Simulation Towards Effective Supply Chain—A Case Study. In Proceedings of the International Conference Management, Penang, Malaysia, 13–14 June 2011.
19. Ibarahim, H.R.; Thani, M.I.; Hussin, R.; Ramlah, W.; Sulaiman, M.S. *Industrial Processes & The Environment (Handbook No. 3): The Crude Palm Oil Industry*; Department of Environment: Putrajaya, Malaysia, 1999.
20. Malaysian Palm Oil Board (MPOB). *Oil Palm Estates, January–December 2017*; MPOB: Kuching, Malaysia, 2018.
21. Sharifzadeh, M. Integration of Process Design and Control: A review. *Chem. Eng. Res. Des.* **2013**, *91*, 2515–2549. [CrossRef]
22. Azman, I. The Impact of Palm Oil Mills' Capacity on Technical Efficiency of Palm Oil Millers in Malaysia. *Oil Palm Ind. Econ. J.* **2014**, *14*, 35–41.
23. Leontief, W.W. Quantitative Input and Output Relations in the Economic Systems of the United States. *Rev. Econ. Stat.* **1936**, *18*, 105–125. [CrossRef]
24. Noel, V.M.; Omega, R.S.; Masbad, J.G.; Ocampo, L. Supply input-output economics in process prioritisation of interdependent manufacturing systems. *Int. J. Manag. Decis. Mak.* **2017**, *16*, 1–23. [CrossRef]

25. Duchin, F. Industrial Input-output Analysis: Implications for Industrial Ecology. *Proc. Natl. Acad. Sci. USA* **1992**, *89*, 851–855. [CrossRef] [PubMed]

26. Ukidwe, N.U.; Bakshi, B.R. Resource Intensities of Chemical Industry Sectors in the United States via Input–Output Network Models. *Comput. Chem. Eng.* **2008**, *32*, 2050–2064. [CrossRef]

27. Egilmez, G.; Kucukvar, M.; Tatari, O.; Bhutta, M.K.S. Supply Chain Sustainability Assessment of the U.S. Food Manufacturing Sectors: A Life Cycle-Based Frontier Approach. *Resour. Conserv. Recycl.* **2014**, *82*, 8–20. [CrossRef]

28. Rocco, M.V.; Di Lucchio, A.; Colombo, E. Exergy Life Cycle Assessment of electricity production from Waste-to-Energy technology: A Hybrid Input-Output approach. *Appl. Energy* **2017**, *194*, 832–844. [CrossRef]

29. Pairotti, M.B.; Cerutti, A.K.; Martini, F.; Vesce, E.; Padovan, D.; Beltramo, R. Energy Consumption and GHG Emission of the Mediterranean Diet: A Systemic Assessment using a Hybrid LCA-IO Method. *J. Clean. Prod.* **2015**, *103*, 507–516. [CrossRef]

30. Tan, R.R.; Aviso, K.B.; Cayamanda, C.D.; Chiu, A.S.F.; Promentilla, M.A.B.; Ubando, A.T.; Yu, K.D.S. A Fuzzy Linear Programming Enterprise Input–Output Model for Optimal Crisis Operations in Industrial Complexes. *Int. J. Prod. Econ.* **2016**, *181*, 410–418. [CrossRef]

31. Jing, D.; Chun-you, W. Multiobjective Optimization Model for Industrial Ecosystem Based on Input-Output Analysis: A Case Study of Combined Heat and Power Plant Eco-industrial Park. In Proceedings of the 2007 International Conference on Management Science and Engineering, Harbin, China, 20–22 August 2007; pp. 1316–1321.

32. Ubando, A.T.; Culaba, A.B.; Tan, R.R.; Ng, D.K.S. A Systematic Approach for Optimization of an Algal Biorefinery Using Fuzzy Linear Programming. *Comput. Aided Chem. Eng.* **2012**, *31*, 805–809.

33. Tan, R.R.; Yu, K.D.S.; Aviso, K.B.; Promentilla, M.A.B. Input–Output Modeling Approach to Sustainable Systems Engineering. *Encycl. Sustain. Technol.* **2017**, 519–523. [CrossRef]

34. Aviso, K.B.; Mayol, A.P.; Promentilla, M.A.B.; Santos, J.R.; Tan, R.R.; Ubando, A.T.; Yu, K.D.S. Allocating Human Resources in Organizations Operating under Crisis Conditions: A Fuzzy Input-Output Optimization Modeling Framework. *Resour. Conserv. Recycl.* **2018**, *128*, 250–258. [CrossRef]

35. Foong, S.Z.Y.; Goh, C.K.M.; Supramaniam, C.V.; Ng, D.K.S. Input–output Optimisation Model for Sustainable Oil Palm Plantation Development. *Sustain. Prod. Consum.* **2019**, *17*, 31–46. [CrossRef]

36. Cleveland, W.S.; Mcgill, R. We have studied the elementary Graphical Perception and Graphical Methods for Analyzing Scientific Data. *Science* **2009**, *229*, 828–833. [CrossRef]

37. Linnhoff, B.; Townsend, D.W.; Boland, D.; Hewitt, G.F.; Thomas, B.E.A.; Guy, A.R.; Marsland, R.H. *User Guide on Process Integration for the Efficient Use of Energy*, 1st ed.; Institution of Chemical Engineers: Rugby, UK, 1982; ISBN 0852951566.

38. Friedler, F.; Tarjan, K.; Huang, Y.W.; Fan, L.T. Graph-theoretic approach to process synthesis: Polynomial algorithm for maximal structure generation. *Comput. Chem. Eng.* **1993**, *17*, 929–942. [CrossRef]

39. Linnhoff, B. Pinch Analysis—A State-of-the-Art Overview. *Chem. Eng. Res. Des.* **1993**, *71*, 503–522.

40. Foo, D.C.Y. State-of-the-art review of pinch analysis techniques for Water network synthesis. *Ind. Eng. Chem. Res.* **2009**, *48*, 5125–5159. [CrossRef]

41. Teng, W.C.; Fong, K.L.; Shenkar, D.; Wilson, J.A.; Foo, D.C.Y. Piper diagram—A novel visualisation tool for process design. *Chem. Eng. Res. Des.* **2016**, *112*, 132–145. [CrossRef]

42. Andiappan, V.; Ng, D.K.S.; Tan, R.R. Design Operability and Retrofit Analysis (DORA) framework for energy systems. *Energy* **2017**, *134*, 1038–1052. [CrossRef]

43. Kasivisvanathan, H.; Tan, R.R.; Ng, D.K.S.; Abdul Aziz, M.K.; Foo, D.C.Y. Heuristic framework for the debottlenecking of a palm oil-based integrated biorefinery. *Chem. Eng. Res. Des.* **2014**, *92*, 2071–2082. [CrossRef]

44. Grossmann, I.E.; Halemane, K.P.; Swaney, R.E. Optimization strategies for flexible chemical processes. *Comput. Chem. Eng.* **1983**, *7*, 439–462. [CrossRef]

45. LINDO Systems Inc. *LINGO the Modelling Language and Optimizer*; LINDO Systems Inc.: Chicago, IL, USA, 2017.

processes

MDPI

Article

Integration of Process Modeling, Design, and Optimization with an Experimental Study of a Solar-Driven Humidification and Dehumidification Desalination System

Mohammed Alghamdi [1,*], Faissal Abdel-Hady [1] , A. K. Mazher [2] and Abdulrahim Alzahrani [1]

[1] Chemical and Materials Engineering Department, King Abdulaziz University, Jeddah 21589, Saudi Arabia; faissalhady@gmail.com (F.A.-H.); azahrani@kau.edu.sa (A.A.)
[2] Nuclear Engineering Department, King Abdulaziz University, Jeddah 21589, Saudi Arabia; amazher@hotmail.com
* Correspondence: mohammed_moghram@hotmail.com; Tel.: +966-50-496-5899

Received: 23 July 2018; Accepted: 5 September 2018; Published: 7 September 2018

Abstract: Solar energy is becoming a promising source of heat and power for electrical generation and desalination plants. In this work, an integrated study of modeling, optimization, and experimental work is undertaken for a parabolic trough concentrator combined with a humidification and dehumidification desalination unit. The objective is to study the design performance and economic feasibility of a solar-driven desalination system. The design involves the circulation of a closed loop of synthetic blend motor oil in the concentrators and the desalination unit heat input section. The air circulation in the humidification and dehumidification unit operates in a closed loop, where the circulating water runs during the daytime and requires only makeup feed water to maintain the humidifier water level. Energy losses are reduced by minimizing the waste of treated streams. The process is environmentally friendly, since no significant chemical treatment is required. Design, construction, and operation are performed, and the system is analyzed at different circulating oil and air flow rates to obtain the optimum operating conditions. A case study in Saudi Arabia is carried out. The study reveals unit capability of producing 24.31 kg/day at a circulating air rate of 0.0631 kg/s and oil circulation rate of 0.0983 kg/s. The tradeoff between productivity, gain output ratio, and production cost revealed a unit cost of 12.54 US$/m^3. The impact of the circulating water temperature has been tracked and shown to positively influence the process productivity. At a high productivity rate, the humidifier efficiency was found to be 69.1%, and the thermal efficiency was determined to be 82.94%. The efficiency of the parabolic trough collectors improved with the closed loop oil circulation, and the highest performance was achieved from noon until 14:00 p.m.

Keywords: desalination; humidification; dehumidification; design; experimental

1. Introduction

The vast majority of the water on Earth is salt water and surrounds available lands. The natural freshwater resources are limited, especially over a stressed water area in the Arabian Peninsula, where renewable energy sources, such as solar energy, are in the highest range. The most common desalination techniques are of a commercial size and designed for large cities, and small populations in villages experience a lack of freshwater availability [1]. Saudi Arabia is located in the Middle East, which has various energy sources, including oil, natural gas, wind and solar radiation [2]. The country is considered one of the largest oil producers in the world, and this industry requires further investment for environmental control in order to reduce the emissions, which contribute to the global warming issue. Therefore, renewable energy sources are more favorable research topics for the promotion of investment in clean technologies, and solar energy is a promising field of research in this connection. The average annual horizontal irradiation in the country is 2200 kwh/m^2/year (251.14 W/m^2) [2,3], which is in the high range compared to different locations around the world. Saudi Arabia's future expectation of solar power demand would be about 30% by the year 2032 [4]. The average annual horizontal irradiance in Saudi Arabia is shown in Figure 1.

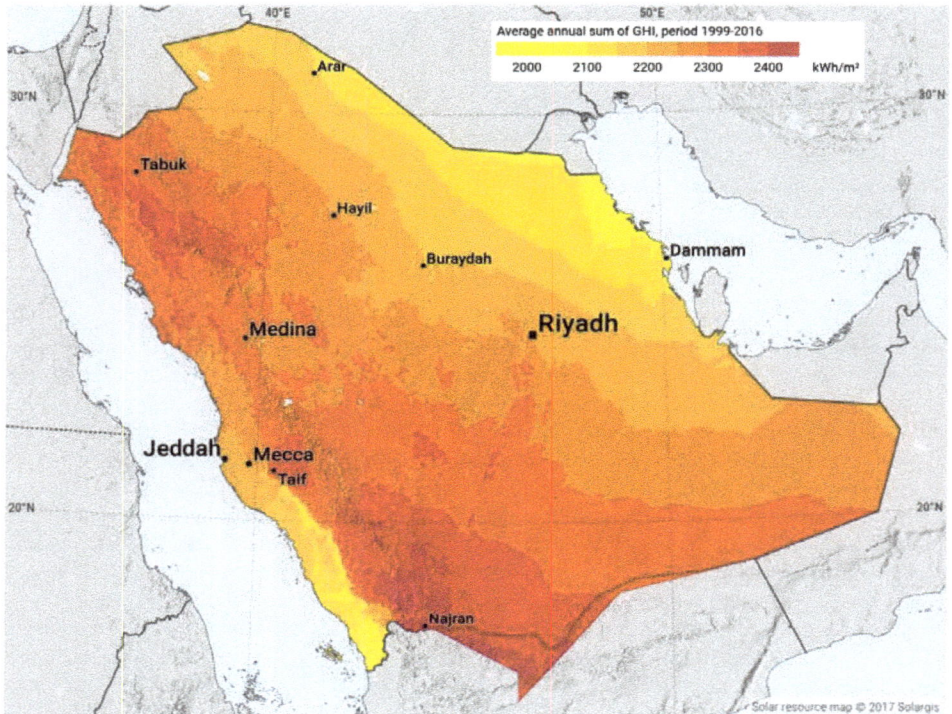

Figure 1. Average annual horizontal irradiation in Saudi Arabia. Reproduced with permission from the solar resource map © 2018 Solargis [5].

Solar energy can be collected using parabolic trough collectors (PTC), which have an efficiency of over 75% [6]. The designed system in this work uses a type SAE 10W-30 synthetic blend engine oil, which circulates through the PTCs to transport the thermal energy to the humidifier heat exchanger; hence, oil flow rates are tested for system performance analysis. The PTCs show higher efficiencies when the circulating fluid rate is increased [7] up to an optimum value, at which a further increase can reduce the efficiency. The flow variation inside the absorber tube is performed [8] in order to relate the Reynolds number to the flow pattern, and the results were obtained in March, April, and May, with a laminar flow. In a review study [9], the PTC performance was found to be dependent on the optical efficiency, circulating fluid heat transfer coefficient, heat flux, reflectance of the reflector, the absorptance, the diameter of the absorber and the length of the collector. The PTC has been used in the desalination field due to its superior performance in solar light tracking and focal receiving [10]. The parabolic concentrator tracking system is used [11] with the tubular solar still for the purpose of desalination, and the resultant production capacities are 0.28, 0.214 and 1.66 L/day, with a production cost of 0.033, 0.044 and 0.024 US$/L, respectively. Moreover, variable pressure humidification and dehumidification (HDH) is used in solar heat in different configurations [12], and the resultant operating costs were 0.034 and 0.041 US$/L, respectively. The PTC is usually installed where the tracking system is accommodated to enhance the performance of the collector. In a study of PTC, with and without a tracking system coupled with a double slope solar still, the tracking PTC is shown to have a higher still temperature and productivity compared with a non-tracked PTC and conventional solar still [13]. The solar PTC performance varies with different latitude angles in different seasons; hence, the desalination performance shows an increase during the day [14]. A solar driven desalination process can be implemented in arid areas, where large commercial technologies are not economically feasible for a small population; hence, the humidification and dehumidification process can be implemented for this purpose. The sophisticated commercial plants are inadequate in arid areas due to the infrastructure requirements, such as sophisticated switchgears, a power grid, fue transportation and pipelines for small production quantities. Moreover, they require a large number of employees, high technical level and maintenance programs; therefore, the solar HDH becomes a promising process for providing freshwater in arid areas with fewer complications associated with serving a small population village. The HDH process operates at lower temperatures and pressures compared to other desalination techniques, which is advantageous in solar desalination practice. Various arrangements of HDH processes can be implemented, and several studies have been performed in this field in order to improve the process outcome and economics [15]. The air flow rate and temperature are crucial factors in the process, where heating the air supply can increase the specific humidity by 1.22 times compared to the ambient air [16]. The gained output ratio (GOR) increases with an increasing air to water ratio. This increase can be achieved by changing either air or circulating water flow rates, and it may be effective up to an optimal value, at which a further increase in the air to water ratio can reduce the system GOR [17]. Hamed et al. [18] concluded that the solar HDH desalination was strongly affected by ambient conditions and, therefore, that the system temperature rise positively affected the productivity from noon until 5 p.m., with the maximum productivity of 22 L/day and production cost of 0.0578 US$/L. Other parameters, such as the cooling water flow and temperature can alter the system productivity. Ahmed et al. [19] found that the HDH inlet cooling water temperature increases the unit productivity by 10 to 15 L/h, with a production cost of 0.01 US$/L. The solar HDH desalination could improve by understanding the effects of configuration and flow circulation rates on the optimum operating conditions. Combining the HDH desalination unit with a renewable-energy source is the target for sustainability and a clean development process. Solar desalination using collectors is still expensive compared to fossil fuels and needs an improvement in order to convert the solar energy to electrical or thermal energy [20]. In order to investigate its process efficiency and economic viability, the process should be analyzed separately before hybridization with another energy source, such as the available fossil fuel in Saudi Arabia [21]. Based on a review study by Kabeel et al. [15], further HDH design simulation, to understand the effects of air and water flow rates and optimal conditions, was recommended. The use of series PTCs, using the motor oil as an alternative to thermal fluids, and coupled to closed air with the recirculated water HDH system, is a promising design for the research field. This can enhance the desalination options for decentralized areas.

This project was implemented at King Abdul-Aziz University at Jeddah city to study the effect of various factors in the solar-driven HDH desalination unit using six PTCs in a series configuration. The system comprises multiple circulation loops in the solar heating cycle and the desalination air and water cycles. This study aimed to extend knowledge about solar energy utilization in the field of clean environment technologies and provide headway to overcome the freshwater demand in areas in which natural water resources are scarce.

2. Experimental Setup and Procedure

The experiments were performed at the King Abdul-Aziz university campus in an open area to allow for the required incidence from the sun at sunrise and sunset in order to achieve a better project analysis and evaluation of its performance. The location latitude is 21 degrees, 30 min north, and the longitude is 39 degrees, 15 min east.

2.1. Experimental Setup

The system is designed to utilize solar energy through multiple circulation procedures. The heat input section circulates the oil through the system of the parabolic concentrator, humidifier, bottom heat exchanger, and the oil expansion tank. The water circulates inside the humidifier in order to maintain maximum water temperature and prevent heat loss, which usually occurs in the open water cycle. On the humidifier-packing surface, the water vapor as well as the latent and sensible heat are transferred to the flowing air using a forced circulation technique, which helps to prevent the loss of air and to increase its temperature and humidity. For such a system, the circulating air mostly operates near the saturation curve and close to the equilibrium condition of air to water, therefore improving the humidifier efficiency. The experimental setup for testing the solar humidification and dehumidification desalination system is illustrated in Figure 2. The parabolic concentrators are used to increase the circulating oil temperature, in which the solar tracking system is used to maximize the beam radiation utilization by increasing the useful heat gain of the concentrators. The oil is driven by a gear oil pump, which provides a proper circulation of oil through the PTCs. Therefore, it delivers the gained energy to the heat exchanger on the bottom of the humidifier to circulate the water heating and allows for the oil thermal expansion in the oil tank. The oil circulation procedure allows the system to build up energy instead of using open circuit designs. The quantity of circulating oil in the tank would provide both stability and possible heat storage, since the circulating oil preserves the heat and requires time to complete the circulation cycle. In order to increase the humidifier inlet water temperature and to deliver the necessary sensible heat to the air and water interface, the water circulates through the packing of the humidifier. The water circulation procedure ensures both the heating and humidifying of air, gaining its energy from the heat exchanger on the bottom of the humidifier. The temperature of the circulating air is increased, and the air contains water vapor to deliver to the dehumidifier. In the dehumidifier, the cooling and condensing of the air and water vapor mixture occurs, and part of the water is extracted from the process as a distillate.

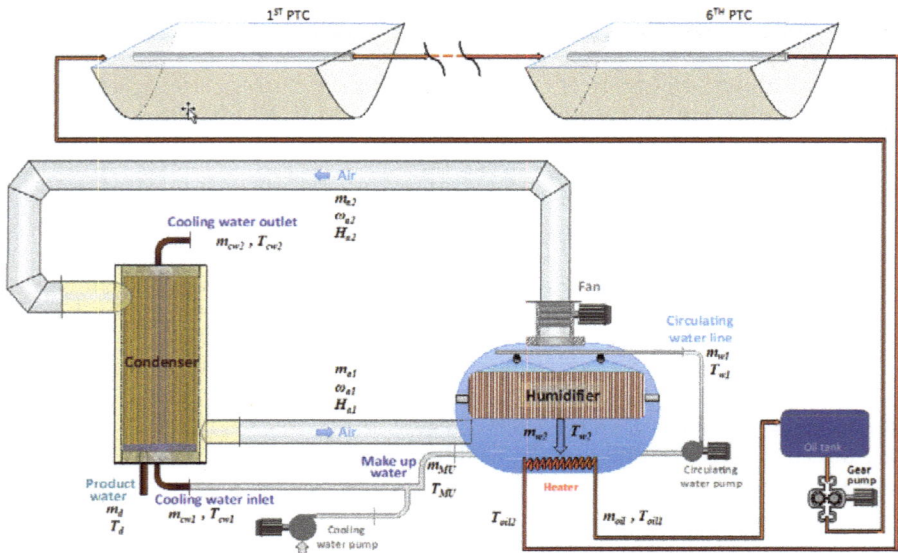

Figure 2. Schematic diagram of the humidifier.

In the dehumidifier the air temperature and its humidity decreased to a point higher than the ambient conditions; therefore, the circulation of this stream allows for further process efficiency improvement. The humidifier circulating water is fed by making water to compensate for the amount of water vapor being transferred to the dehumidifier and is extracted as a process yield. The fabricated and installed major components of the experiment are shown in Figure 3.

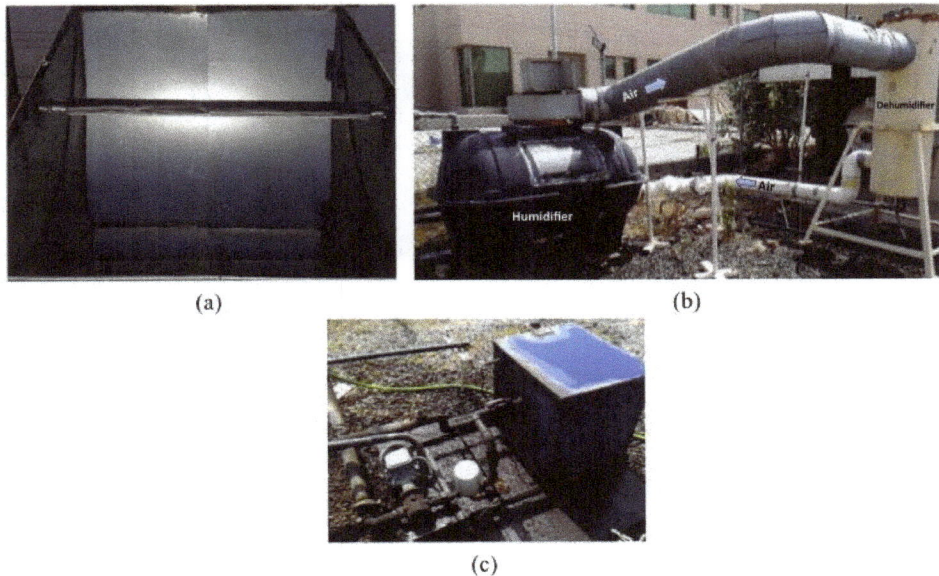

Figure 3. Photos of the experimental setup and main components: (**a**) parabolic trough collector (PTC) with absorber tube; (**b**) humidification and dehumidification (HDH) desalination unit; and (**c**) oil expansion tank.

2.1.1. The Parabolic Concentrators

There are six parabolic concentrators aligned in a series. The solar beam radiation reflects from the concentrator on the lower part of the absorber tube. The six parabolic concentrators are made from glass fiber-reinforced polymer (GFRP), with a thickness of 2.3 mm. The GFRP material was selected due to its availability and cost-effectiveness and, moreover, because it is known to have an excellent thermal insulation. The GFRP are lined with a polymer-based mirror type 3M fused to a stainless steel sheet, with a thickness of 0.5 mm. The polymeric mirror allows the concentrator to have a high efficiency at the working incidence. The dimensions of one PTC used in this experiment are shown in Figure 4. The absorber tube is made from a copper material, which has a high thermal conductivity and is lightweight. The tube is coated with black polymeric paint, which allows the receiver tube to absorb the radiation more efficiently.

The design specification of the PTC and the fabrication material are listed in Table 1. The tracking system of the PTC is designed to work from east to west while the PTCs are oriented in the horizontal north–south direction.

A thermally-treated protective glass covers the aperture plane, which aids in the reduction of the convective thermal loss due to winds flowing around the absorber tube. The solar beam radiation, measured as normal, on the PTC surface during the experiment is recorded by a precise solar power meter. The meter is an instrument, used in the solar research field, of type SM206, with an error range of ±5% of the measured value.

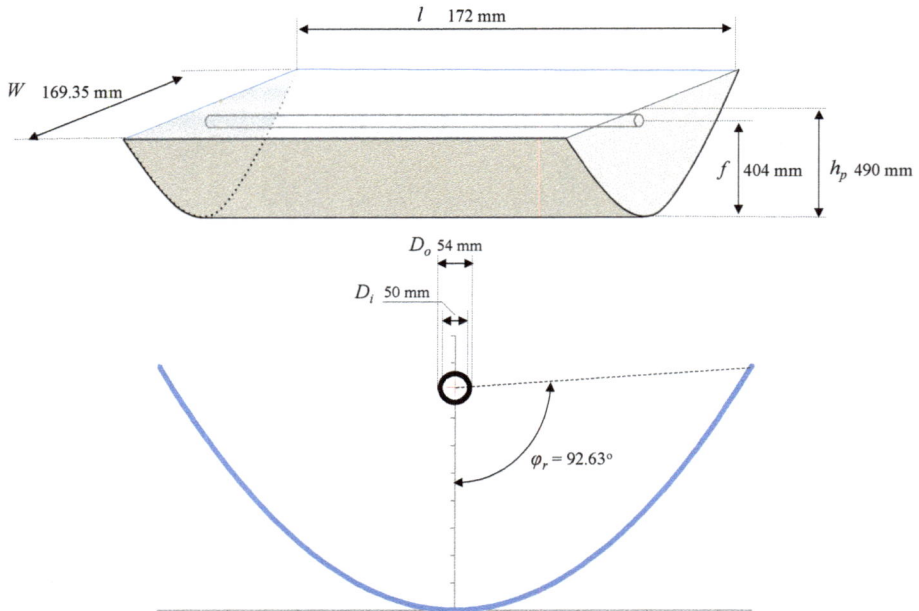

Figure 4. Schematic diagram of the parabolic concentrator.

Table 1. The PTC geometrical parameters.

r_r	Rim radius, m	0.848
φ_r	Rim angle	92.63°
f	Parabola focal distance, m	0.404
D_o	Receiver outside diameter, m	0.054
D_i	Receiver internal diameter, m	0.05
θ_m	Acceptance angle	2.84°
W_a	The aperture width, m	1.694
h_p	Height of the parabola, m	0.49
l	Trough length, m	1.72
C	Geometric concentration ratio	9.89
s	Curve length of the reflective surface, m	1.966
t	Glass thickness, m	0.004
J	Reflector thickness, m	0.001
ξ	Reflectance of the reflector	0.95
τ	Transmittance of the glass cover	0.9
ψ	Absorptance of the receiver.	0.97
γ	Intercept factor	1.000

Engine oil is used as the working fluid in this experiment due to its durability and lower price compared to commercial thermal fluids. The oil type is SAE 10W-30 synthetic blend engine oil. Engine oil has an advantage over water, since oil can sustain higher temperatures unlike water, which evaporates and develops hammering in the pipes, requiring high-quality piping and joint fabrication, and this increases the process capital cost. The thermo-physical properties of the circulating oil can change when its temperature increases during solar heating. The thermo-physical properties, shown in Table 2, are used to proceed with the system calculations.

Table 2. Thermo-physical properties of the SAE 10W-30 synthetic blend engine oil.

T Temp. (K)	ρ Density (kg/m^3)	c_p Specific Heat (kJ/Kg-c)	μ Viscosity (N-s/m^2)	ν Kinematic Viscosity (10^{-4} m^2/s)	k Thermal Conductivity (W/m-K)
260	908	1.76	12.23	135	0.149
280	896	1.83	2.17	24.2	0.146
300	884	1.91	0.486	5.50	0.144
320	872	1.99	0.141	1.62	0.141
340	860	2.08	0.053	0.62	0.139
360	848	2.16	0.025	0.30	0.137
380	836	2.25	0.014	0.17	0.136
400	824	2.34	0.009	0.11	0.134

2.1.2. Humidification and Dehumidification Unit

The humidifier tank is made from fiberglass-reinforced plastic (FRP) material that has the following advantages: it is lightweight, corrosion resistant, cost-effective and robust enough to handle the system working pressure and temperature. There are different sizes available for such a tank, which can be selected upon the operating capacities of the desalination unit. Moreover, the humidifier body is painted with a black polymer coating to enhance the sunbeam radiation absorption during the day, when the temperature of the system, designed to promote the circulating water, increases. The FRP material is known to have a low thermal conductivity; hence, the combination of both FRB and black paint will enhance the energy conservation in the humidifier. The opening on the top of the humidifier is equipped with a forced draft fan, driven by a 220 v 3-phase motor, to circulate the humid air through the system. A schematic diagram of the humidifier components, dimensions, and flow openings is shown in Figure 5. Several ports are fabricated for the returned air flow from the dehumidifier, makeup water line, circulating water inlet and oil circulation through the bottom heater. The humidifier is internally packed with canvas sheets to promote air-to-water contact by increasing the surface area due to its high porosity. This material is selected for its common availability, low cost and easy replacement, which reduces both capital and running costs. The packing material is supported in the middle of the humidifier tank using a transparent acrylic sheet manufactured from a synthetic polymer. The acrylic sheet is also used to seal and prevent air from passing around the packing in a shortcut, away from the packing. It is installed using adhesives around the middle edge of the humidifier, where the packing is placed. The circulating water is introduced into the humidifier through two top nozzles to ensure a conical shape of flow covering the packing surface area. Both nozzles are capable of delivering a circulating water pump capacity of 0.571 kg/s. The circulating air and water are flowing in a counter flow pattern inside the humidifier. The circulating water is a forced type, using a single-phase constant speed pump driven by a 220 v motor. The bottom heater of the humidifier is an unmixed counter flow heat exchanger, typically used in the internal combustion engine cooling system of automobiles. The heat exchanger manufacturing material is aluminum, with a total surface area of 11.074 m^2. The heating oil flows inside the tubes are crosswise to the circulating water direction, where the water is maintained above the heat exchanger using a float valve in the makeup water inlet port.

The dehumidifier is a vertically mounted shell and tube heat exchanger, with one pass and no baffles. The humid air stream is introduced into the dehumidifier from the top and flows downward to the bottom of the heat exchanger, and it is then released from the outlet port and recycled for the humidifier. A schematic diagram of the dehumidifier shell and tube heat exchanger is shown in Figure 6. The design of the implemented heat exchanger uses 50 copper tubes, each with an outer diameter of 28.5 mm. The height of the copper tubes is 1300 mm, which together provide a 5.817 m^2 contact surface area between the humid air and the cooling water. There are two water boxes for the heat exchanger, which are made from galvanized steel, with a height of 50 mm and 470 mm in diameter, providing 0.0174 m^3 of cooling water for both water boxes. The cooling water flows from the bottom water box upward through the copper tubes and is then released from the top outlet pipe to ensure that the tubes are uniformly filled, even at lower water cooling flow rates. The humid air flows counter-current to the cooling water stream, in which part of the entrained water vapor condenses at the outer surface of the copper tubes and is then diverted to the distillate water port for collection.

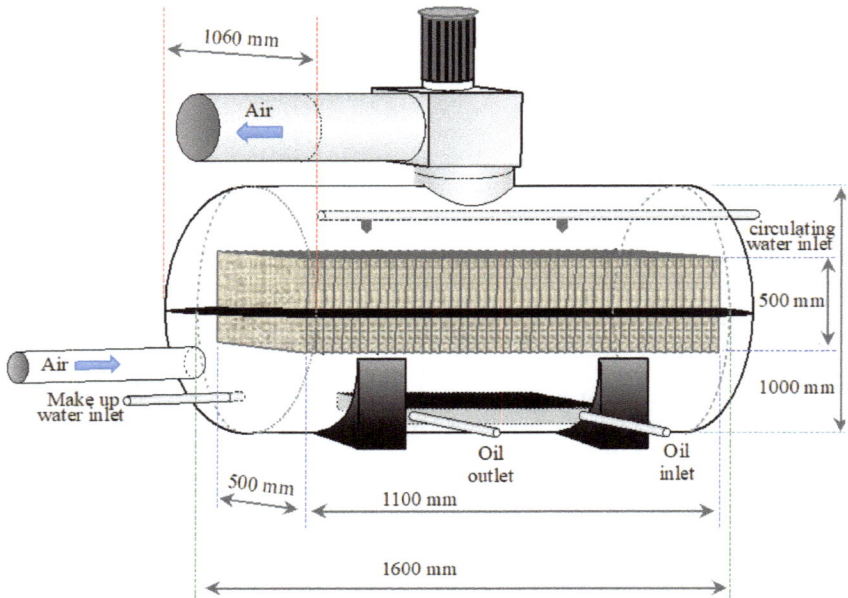

Figure 5. Schematic diagram of the humidifier.

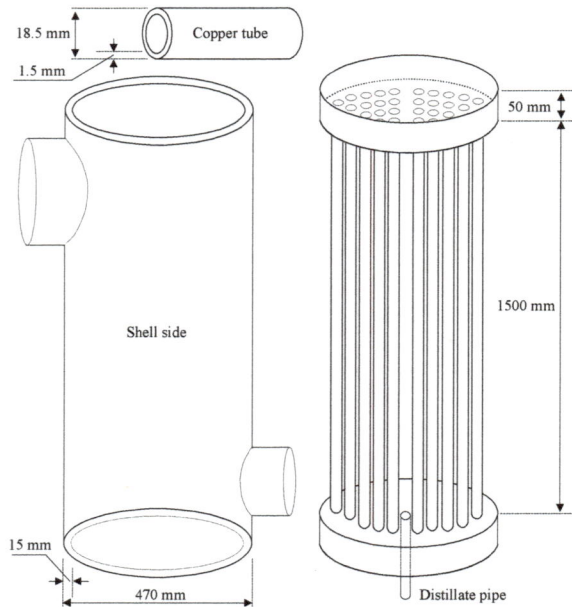

Figure 6. Schematic diagram of the dehumidifier.

2.1.3. Measuring and Controls

Both the desalination unit and the solar tracking system of the PTCs are controlled and monitored using Laboratory Virtual Instrument Engineering Workbench (LabVIEW) software. For solar system tracking, the designed graphical interface using a visual programming language is enabled to control the tracking DC motor motion through the amplified signal, sent by a Galil motion control card. A control loop with a feedback signal allows the motion control system to continuously track throughout the day, as shown in Figure 7.

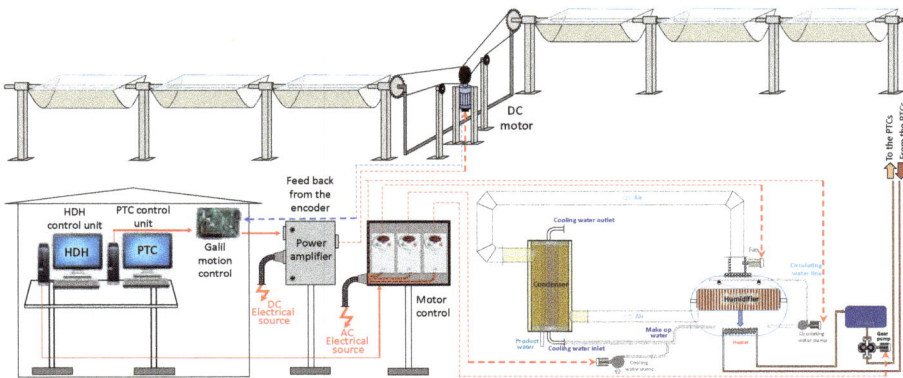

Figure 7. Schematic diagram of the experiment controls.

The east and west rotations of PTCs are protected by limit switches, installed in both directions for controlling the motor operation at sunset, to return the PTCs to the home position and stop the motor when the home position is reached as well as prepare the collector for the sunrise position of the next day. An absolute type encoder, mounted on the DC motor shaft, controls the angular position of the PTCs to monitor the motion activity. The desalination unit drives and the circulating oil pump drive speed are controlled by an AC frequency inverter, which provides a wide range of speed control; hence, the flow capacities are controlled based on the desired testing value. The circulating water pump of the humidifier is kept at a constant speed, while the air circulating fan speed is altered. The system oil/air/water temperatures are measured using K-type thermocouples, which have a wide range of sensing temperatures, with an accepted accuracy at error limits of ±0.75% [22]. The thermocouples are connected to the computer through a field point thermocouple input model. Daytime relative humidity, wind speed, and ambient temperatures are recorded. The circulating air conditions are recorded, and its temperature and humidity are measured by a humidity sensor mounted on the air ducts. The product water from the desalination unit is collected from the distillate port under the dehumidifier and measured in a scaled water storage tank. The measured experimental data uncertainty analysis is described in Appendix A, however, the maximum standard uncertainty is found to be 0.94.

2.2. *Experimental Procedure*

Each experimental operation and recording is performed on a daily basis. The oil expansion tank level is checked in an adequate position. The oil, air, humidifier water, dehumidifier cooling water circulation are started at 06:20 a.m. daily to ensure the system functionality. The circulation provides a uniform temperature over the cycles and complete humidifier packing constitutes a wet condition. The desired oil and air circulating flow rates are established using the AC frequency inverters. The PTCs are checked in the home position, preparing for the sunrise and tracking the solar motion throughout the day. The computer graphical interface of LabVIEW is initiated to start the tracking system using a Galil motion control card. Similarly, the system temperatures and humidity recordings are stored in an excel sheet for the whole experiment run. The solar beam radiation is measured as normal on the horizontal level of the PTCs and is recorded every hour. The distillate obtained from the desalination unit is collected and measured hourly. Sun tracking and system recording are continued until the end of the day, when the return limit switch is activated to rotate the PTCs back to the home position. For the next testing conditions, the circulating oil and air flow rates are changed based on the test plan to record several operation modes in order to analyze the designed system performance.

3. Experiment Analysis

The desalination system can be split into three major control volumes in order to perform analyses and calculations for each one. The control volumes are classified as follows:

(1) Solar parabolic concentrator,
(2) Humidifier,
(3) Dehumidifier.

The direction of the heat flow starts from the solar collector through the humidifier to the dehumidifier, where part of the heat is retained by recirculating the air back to the humidifier, and the insignificant heat loss is dumped into the cooling reservoir.

3.1. Parabolic Solar Collector

3.1.1. Solar Irradiance and Tracking Angles

The beam from the sun can generally be considered as the direct beam and diffused beam; hence, the solar irradiance to the parabolic concentrator would be combined beams. Due to the geometrical design of the parabolic concentrator, the direct beam is considered the useful part of the solar irradiance, and it is affected by the incidence angle and called the tilted beam radiation [23], which can be given by the following relation:

$$I = G_{bn} cos(\theta) \tag{1}$$

where I and G_{bn} are the solar irradiance and the normal beam radiation, respectively. The incidence angle θ must be calculated in order to obtain the useful solar irradiance to the parabolic concentrator and can be found by calculating other solar angles.

- Latitude angle (L)

This depends on the location of the site on the Earth and the site of the experiment found above the equator; hence, the latitude of the experiment site is positively valued and found to be 21.5°.

- Hour angle (h)

Solar noon is defined as that time of day when the sun appears due south (north), and the solar time is the time of day measured from solar noon [24]. The reference day hour is 12:00 p.m., where the mid daylight hours are calculated to have the maximum expected radiation. The hour angles are negatively valued for the first half of the day and positively valued in the afternoon. The following relation is employed to calculate the hour angle [17,19]:

$$h = 15^0 [t_s - 12] \tag{2}$$

where t_s is the solar time based on the local time and substituted in a 24-h system.

- Declination angle (δ)

The declination angel is influenced by the location of the Earth with respect to the sun. The value of this angle varies positively or negatively depending on the season and specifically on the number of days in the year due to the rotation of the Earth. The following relation gives the declination angle [23,25]:

$$\delta = 23.45 \left[sin \left(\frac{360}{365}(284 + N) \right) \right] \tag{3}$$

where N is the number of the day in the year. For the experiment site, the maximum positive declination angle for the earth is 23.45 in the summer, and the maximum negative value is -23.45 in the winter.

- Altitude angel (α)

The altitude angle is calculated for the horizontal surface of the experiment site. It depends on the elevation of the sun in the sky; hence, the projection of the sunbeam on the horizontal surface forms this angle. Therefore, the experimental site latitude, hour angle and declination angle of the earth with respect to the sun have a great influence on this angle. The following relation is used to calculate it [23,26]:

$$\alpha = sin^{-1}[sin(L)sin(\delta) + cos(L)cos(\delta)cos(h)] \tag{4}$$

- Azimuth angle (ϕ)

The azimuth angle is concerned with the south direction of the experimental site. It is the angle between the south direction of the experimental site and the sunbeam projection on the horizontal site and can be calculated as follows [23]:

$$\phi = sin^{-1} \left[\frac{cos(\delta)sin(h)}{cos(\alpha)} \right] \tag{5}$$

- Surface azimuth angle (ϕ_s)

The surface azimuth angle is the deviation of the projection, on a normal horizontal plane on the surface, from the local meridian [27]. For the present case, where the tracking system operates from east to west, this angle will be either $-90°$, when the concentrator is facing east, or $90°$, when it is facing west [23].

- Zenith angle (θ_z)

The magnitude of the zenith angle is obtained from the difference between the altitude angles and the vertical position on the surface, which depends on the sun's position in the sky. The zenith angle is calculated by the following relation [28]:

$$\theta_z = 90° - \alpha \tag{6}$$

- The incident angle (θ)

This is the sunbeam projection angle and is measured from the vertical axis to the surface. It is obtained by Kreith and Kreider, and Duffie and Beckman [23,27,29,30], and is shown as follows:

$$\theta = cos^{-1} \left[\sqrt{sin^2(\alpha) + cos^2(\delta)sin^2(h)} \right] \tag{7}$$

3.1.2. Concentrator Thermal Analysis

The optical efficiency of the solar parabolic concentrator is required in order to proceed with the calculation of the thermal efficiency of the system. It represents the percentage of the absorbed solar irradiance "energy" that falls on the absorber tube. This efficiency accounts for the system's random and non-random errors. The equation used to calculate the optical efficiency is given by [23,27]:

$$\eta_o = \xi \tau \psi \gamma \left[\left(1 - A_f tan(\theta) \right) cos(\theta) \right] \tag{8}$$

where A_f is the geometric ratio between the aperture lost area A_l of the concentrator and the total aperture area A_a and can be obtained by the following relation:

$$A_f = \frac{A_l}{A_a} = \frac{\frac{2}{3}Wh_p + fW\left[1 + \frac{W^2}{48f^2}\right]}{Wl} \tag{9}$$

The geometric ratio of the design parabolic concentrator is calculated and found to be A_f = 0.511. The intercept factor γ for the concentrator is obtained by Abdel-Hady F. et al. [31] and found to be 0.84, when the absorber tube outer diameter is 0.42 mm, and still 16.3% of rays dissipated away from the absorber tube. Therefore, the absorber tube diameter is enlarged to 0.54 mm to allow for maximum sunbeam coverage; hence, the intercept factor becomes 1, and the collector concentration ratio $C = W/\pi D_o$ is decreased to 9.984. The lost area, due to the end effect, is not an effective area and is significantly affected by the solar incidence angle θ, which can be calculated by the following relation [23]:

$$A_e = fWtan(\theta)\left[1 + \frac{W^2}{48f^2}\right] \tag{10}$$

The concentrator instantaneous efficiency is known to be the ratio of the amount of heat gained to the amount of solar irradiance that falls on the absorber tube within its effective surface area. The following relation is used to calculate this efficiency [23,27,28]:

$$\eta = \frac{m_{oil}c_{oil}\left(T_{c-oil-in} - T_{c-oil-out}\right)}{I(A_a - A_e)} \tag{11}$$

The heat removal factor F_R, in Equation (13) is employed to calculate the collector instantaneous efficiency, which is based on the concentrator fluid inlet temperature T_i and ambient temperature [28,30].

$$\eta = F_R\left[\eta_o - \frac{U_L}{C}\left(\frac{T_i - T_a}{I}\right)\right] \tag{12}$$

From the experimental data, obtained daily at different operating conditions, the required variables are measured and analyzed. ASHRAE Standard 93-2003, "Methods of testing to determine the thermal performance of solar collectors" (ASHRAE 2003), is used. The concentrator instantaneous efficiency η is calculated from Equation (12), where the term $(T_i - T_a)/I$ in Equation (13) is measured experimentally. The plot of the instantaneous efficiency against $(T_i - T_a)/I$ reveals the value of $F_R \eta_o$, which represents the intercept of the line, and $(F_R U_L/C)$ is the slope of this line; hence, the values of F_R and U_L are obtained. The overall heat transfer coefficient of the absorber tube can be calculated from the following relation [23,28]:

$$U_c = \left[\frac{1}{U_L} + \frac{D_o}{h_{ab}D_i} + \frac{D_o ln\left(\frac{D_o}{D_i}\right)}{2k_{ab}}\right]^{-1} \tag{13}$$

where the convective heat transfer coefficient h_{ab} inside the absorbers tube is calculated from the following relation [32]:

$$Nu = 0.023Re^{0.8}Pr^{0.4} \tag{14}$$

For the fully developed laminar flow pattern, the Nu number value is 4.36 [32]. The Re and Pr numbers are obtained from the following relations, using the fluid properties inside the absorber tube.

$$Re = \frac{4m_{oil}}{\pi D_i \mu_{oil}} \tag{15}$$

$$Pr \frac{c_{oil} \mu_{oil}}{k_{oil}} \tag{16}$$

3.2. Humidifier

3.2.1. Humidifier Bottom Heater Section

A cross flow type heat exchanger, in which the oil flows inside the tubes in a cross direction to the outlet water from the humidifier packing, which flows downward through the heat exchanger, is incorporated in the humidifier. The heat exchanger design parameters and the experimental data are used to calculate the heat exchanger overall heat transfer coefficient in the following manner. The effectiveness relation of the cross-flow heat exchanger, with a single pass and both fluids unmixed, are given below [32,33]:

$$\varepsilon = 1 - exp\left[\left(\frac{1}{C_r}\right)(NTU)^{0.22}\left(exp\left[1 - C_r(NTU)^{0.78}\right] - 1\right)\right] \tag{17}$$

where C_r is the heat capacity ratio, which is given as $C_r = C_{min}/C_{max}$. The heat capacities are calculated for the cold stream $C_w = m_w c_w$, and the hot fluid $C_{oil} = m_{oil} c_{oil}$ is then categorized based on their values as maximum or minimum.

The value of the heat exchanger effectiveness is calculated from the experiment heat analysis as follows:

$$\varepsilon = \frac{q_{HD,in}}{C_{min}(T_{oil,i} - T_{water,i})} \tag{18}$$

From the previously obtained parameters, Equation (18) is solved for the NTU value at the operating condition. Therefore, NTU is used to calculate the overall heat transfer coefficient of the humidifier bottom heat exchanger as shown below.

$$U_{HD} = \frac{(NTU)C_{min}}{A_{HD_{heater}}} \tag{19}$$

The mass and energy balance across the heater is obtained, based on the configuration of the streams, as shown in Figure 7.

Mass balance:

$$m_{w1} = m_{w2} + m_{MU} \tag{20}$$

Energy balance:

$$m_{oil}c_{oil}(T_{oil2} - T_{oil1}) = m_{w2}c_w T_{w2} + m_{MU}c_w T_{MU} - m_{w1}c_w T_{w1} \tag{21}$$

where

$$m_{MU} = m_a(\omega_{a2} - \omega_{a1}) \tag{22}$$

3.2.2. Humidifier Packing Section

According to the designed humidification and dehumidification desalination unit, shown in Figure 8, the heat and mass balance equations are obtained from the control volume 1, as follows:

Mass balance across the packing:

$$m_{w2} - m_{w1} = m_a(\omega_{a2} - \omega_{a1}) \tag{23}$$

Energy balance for air and water streams:

$$m_a(H_{a2} - H_{a1}) = m_{w1}c_w T_{w1} - m_{w2}c_w T_{w2} \tag{24}$$

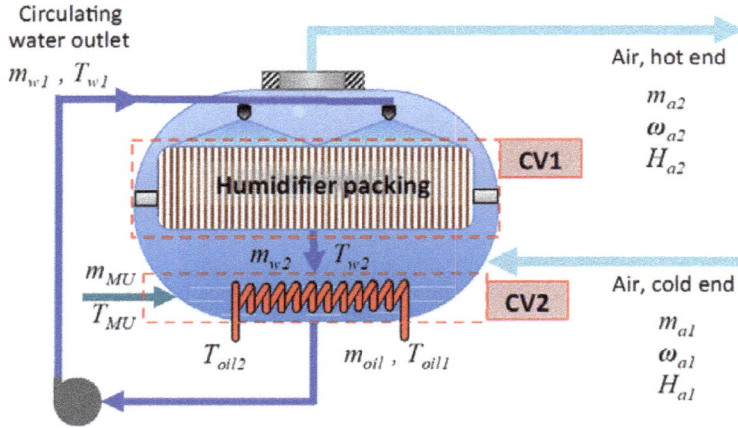

Figure 8. Schematic of the humidifier balance.

The mass transfer coefficient of the humidifier is obtained by using the log mean enthalpy difference (LMED) method. This method showed a good consistency with *ε-NTU* [34,35]; hence, the formulation of the balance is obtained as follows:

$$m_a(H_{a2} - H_{a1}) = K_y aV\Delta H_m \tag{25}$$

where ΔH_m is the log mean enthalpy difference:

$$\Delta H_m = \frac{(H_{as2} - H_{a2}) - (H_{as1} - H_{a1})}{ln\left(\frac{(H_{as2} - H_{a2}) - f_c}{(H_{as1} - H_{a1}) - f_c}\right)} \tag{26}$$

where f is the developed analytically-based correction factor, which is calculated from:

$$f_c = \frac{H_{as1} + H_{as2} - 2H_{avg}}{4} \tag{27}$$

3.3. Dehumidifier

The flow streams of the dehumidifier (CV2) are shown in Figure 9. The heat and mass balance of the dehumidifier are formulated as follows:

Mass balance:

$$m_d = m_a(\omega_{a2} - \omega_{a1}) \tag{28}$$

Energy balance:

$$m_a(H_{a2} - H_{a1}) = m_{cw}c_w(T_{cw2} - T_{cw1})m_d c_w T_d \tag{29}$$

where

$$T_{cw2} = T_d$$

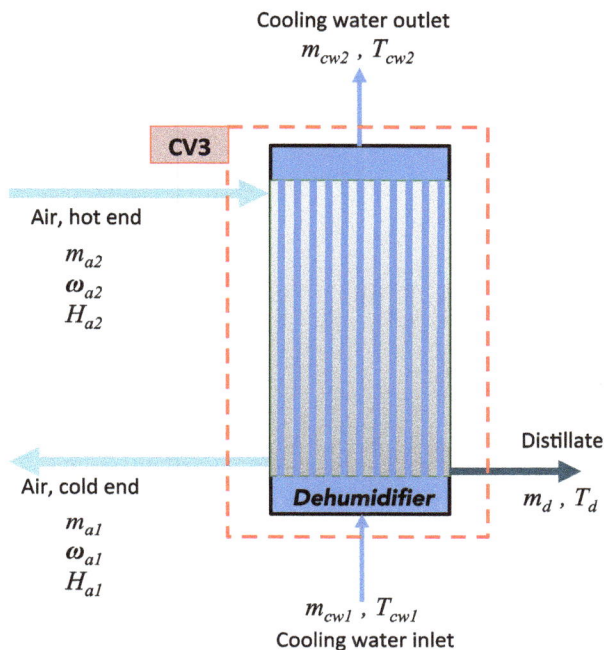

Figure 9. Schematic of the dehumidifier balance.

In order to calculate the overall heat transfer coefficient in the condenser, the log mean enthalpy difference LMED method [32] is used. This method is adequately used in the present case, where the dehumidifier is considered as a shell and tube heat exchanger with a single pass. The applied balance across the condenser is shown below.

$$m_{cw}c_{cw}(T_{cw2} - T_{cw1}) = UA_{DH}\Delta T_{lm} \tag{30}$$

where ΔT_{lm} is calculated as:

$$\Delta T_{lm} = \frac{(T_{a2} - T_{cw2}) - (T_{a1} - T_{cw1})}{ln\left(\frac{T_{a2}-T_{cw2}}{T_{a1}-T_{cw1}}\right)} \tag{31}$$

3.4. System Key Performance Parameters

3.4.1. Humidifier Efficiency

This factor shows the capability of the humidifier to transfer the amount of water vapor, compared to that in the full saturation condition. It indicates the performance of the design, along with the packing, and can be calculated as follows:

$$\eta_H = 100\left(\frac{\omega_{a2} - \omega_{a1}}{\omega_s - \omega_{a1}}\right) \tag{32}$$

The energy-based effectiveness of the humidifier is a non-dimensional value, which can evaluate the humidifier performance based on the maximum possible change in the total enthalpy rate. The energy effectiveness is calculated based on the following [36]:

$$\varepsilon_{HD} = \frac{m_a c_a (T_{a2} - T_{a1})}{C_{min}(T_{a1} - T_{w1})} \tag{33}$$

3.4.2. System Gain Output Ratio (GOR)

The solar energy is used in this system as the heating source. The yield from the desalination is the amount of distillate produced; hence, the yield over the input energy to the system represents the gain output ratio of the system:

$$GOR = \frac{m_d h_{fg}}{Q_{in}} \tag{34}$$

3.4.3. Humidifier Thermal Efficiency

In order to measure the energy conversion in the humidifier, the input energy to the system is related to the output from the humidifier. It indicates how much thermal energy is lost in the humidifier and not being used for the purpose of humidification:

$$\eta_{T_{HD}} = 1 - \frac{Q_{out_{HD}}}{Q_{in}} \tag{35}$$

$$Q_{out_{HD}} = m_a c_a (T_{a2} - T_{a1}) + m_a (\omega_{a2} - \omega_{a1}) c_w T_d \tag{36}$$

3.4.4. Convective Heat and Mass Transfer Coefficients

The ratio of the convective heat transfer coefficient to the convective mass transfer coefficient represents the amount of heat per unit volume required to raise the temperature by one degree. This ratio is affected by the type of process components, insulation and air/water stream flow variations. The ratio is calculated from the Chilton–Colburn analogy [26]:

$$\frac{h_H}{h_M} = \rho_a c_a \left(\frac{Sc}{Pr}\right)^{\frac{2}{3}} \tag{37}$$

where

$$Sc = \frac{\mu_a}{\rho_a D_{water/air}} \tag{38}$$

The diffusivity of water in the air is calculated by the following relation [37] in the temperature range of 273 $K < T < 373$ K:

$$D_{water/air} = 1.97 \times 10^{-5} \left(\frac{T_{a_{avg}}}{256}\right) \tag{39}$$

3.5. Thermodynamic Relations

The properties of the fluids are calculated using the thermodynamic relations. The following relation calculates the absolute humidity [38]:

$$\omega = 0.62 \frac{P_v}{P_T - P_v} \tag{40}$$

The partial pressure P_v of water vapor is calculated from Antoine equation [39]:

$$ln\, P_v = 18.3036 - \frac{3816.44}{227.02 + T(^oC)} \tag{41}$$

The humid air enthalpy is calculated by [38]:

$$H_a = [c_a T_a + 597.2\omega] \times 4186.8 \tag{42}$$

The specific heat of humid air c_a [39] is obtained from the following relation:

$$c_a = [0.24 + 0.46\omega] \tag{43}$$

3.6. Process Economics

The preliminary economic analysis is based on the experiment cost information obtained during the construction process. The capital investment data, including the installation, are collected and shown in Table 3, which can be used for the production cost analysis.

Table 3. Capital cost of the HDH unit.

Description	Cost	
HDH desalination system	2600	US$
Oil tank and circulation system	350	US$
Control unit and sensors	750	US$
Solar concentrator system	5000	US$
HDH Capital cost (*P*)	8700	US$

The summation of fixed and variable operation costs gives the cost of the production, where the total operation cost will include the operation, maintenance, and the annual capital cost for the process, which can be obtained by the following relation [40]:

$$C_{TOP} = C_A + C_{OP} + C_M \tag{43}$$

The maintenance cost is taken as 3% [41,42] of the annual capital charges, which can be less than 2%, as estimated by El-Dessouky et al. [43]. The uniform series of payment method is used to calculate the annual capital cost (C_A). The capital recovery of the project, invested at a 5% interest rate for 20 years, is obtained from the following relation, as an annualized form [44]:

$$Anual\ capital\ cost\ (C_A) = P \frac{[i(1+i)^n]}{[(1+i)^n - 1]} \tag{44}$$

The calculations accounted for the variable operating cost, which is influenced by the price of the energy consumed by the process in the drives and control system. The local energy price is 0.048 US$/kWh [45], as the standard price for the industrial sector in Saudi Arabia. The unit is assumed to operate 10 h per day, with the availability of 90% of the year. The following relation is used to calculate the production cost:

$$Production\ cost = \frac{Total\ production\ cost\ (C_{TOP})}{Annaul\ production} \frac{US\$}{m^3} \tag{45}$$

4. Results and Discussion

The system was operated and tested to evaluate the key performance factors and the productivity at different operating conditions.

4.1. System Productivity

The HDH unit productivity is affected by the circulating air to water ratio as well as by the circulating oil flow rate for the designed solar-driven process. The process has been tested at different air flow rates with a constant oil flow over a certain time period; then, the oil flow rate is changed, and the air flow is set to the next desired testing flow rates. System productivity, at different air and oil flow rates, is illustrated in Figure 10. The objective of increasing the air flow rate is to obtain the highest entrained water in the air at the maximum efficiency of the humidifier in order to separate them by condensation in the dehumidifier section. The maximum obtained system productivity is 24.31 kg/day at the air flow of 0.0631 kg/s. This productivity is enhanced by the oil flow of 0.0983 kg/s, which delivered the thermal energy to the humidifier at the maximum balance between the incoming solar irradiance and the air cooling effect in the humidifier. Moreover, the effect of increasing air flow rate on the productivity, which shows a reduction in the process yield when the air flow rate is increased, is presented in Figure 10.

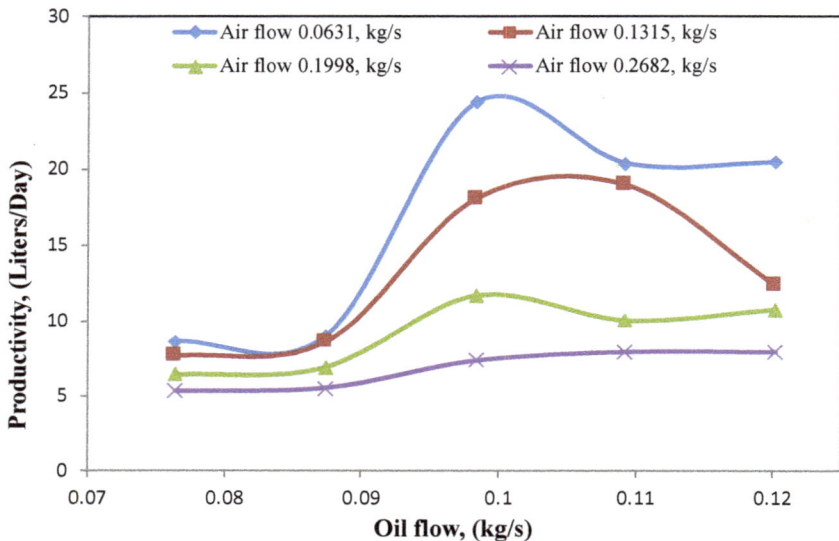

Figure 10. The productivity of the system at different air and oil flow rates.

At a circulating oil flow rate of 0.0983 kg/s, the highest yield is obtained. The productivity of the system is reduced by 26% when the air flow increased to 0.1315 kg/s and further decreased by 52.2% and 70.1% at air flow rates of 0.1998 kg/s and 0.2682 kg/s, respectively. The circulating water temperature has a significant role in the humidifier performance, where the system's highest productivity is obtained at a circulating water temperature of 49.23 °C. The circulating water temperature is decreased when the circulating air flow rate increases, and the maximum reduction was 21.7% at the highest air flow rate. The experimental output is shown to have some fluctuations due to many factors that interfere with the process stability, such as the climate conditions and characteristics of the independent components. In general, each tested condition shows a trend in the process behavior despite the fluctuation rate, which indicates the system outcome. The heat input on the bottom heater of the humidifier provokes the productivity of the system by increasing the humidifier circulating water temperature, which is shown in Figure 11. In the humidifier unit, the highest oil temperature is achieved when the unit performance is combined with the oil flow rate, achieving the required energy input for the maximum circulating water temperature. Increasing the oil flow positively affects the process productivity until the point of the tradeoff between either the installation of an enormous humidifier heat exchanger or a larger solar concentrator aperture area. The reduction of heating at higher oil flow rates is attributed to many factors, such as the absorber heat removal coefficient and less resident time of the oil flows inside the absorber tube and humidifier heat exchanger.

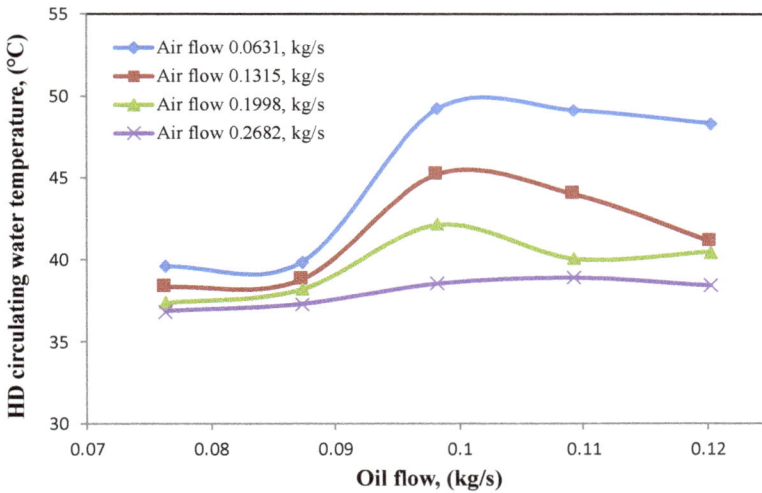

Figure 11. Average HD water temperature at different air and oil flow rates.

4.2. Energy of the System

Based on the oil circulation flow rate, the solar energy gain varies; hence, the input energy to the heat exchanger on the bottom of the humidifier is affected by this variation. The circulating air entering the humidifier has the effect of cooling the system, which has lost the sensible heat in the dehumidifier. The energy input to the heat exchanger on the bottom of the humidifier is illustrated in Figure 12. In the present system, the energy input increased, in most cases, as the oil circulation rate increased up to the desired heating condition, which ranges from 2.39 to 3.07 kWh at an oil flow of 0.0983 kg/s. At a higher humidifier air flow rate, the bottom heat exchanger did not yield the required input energy, and the energy supplied was reduced by 39.3% at the maximum oil flow rate compared to the highest obtained air flow rate of 0.0631 kg/s.

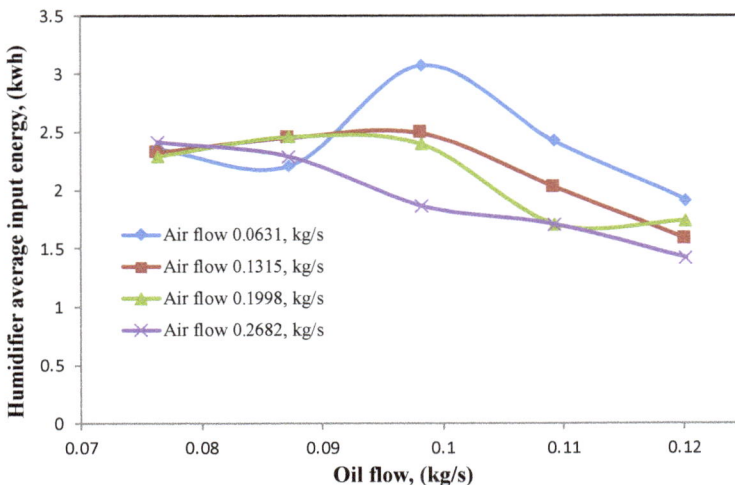

Figure 12. HD average input energy at different air and oil flow rates.

4.3. The GOR

At the low oil flow rate, the process revealed no significance in the unit productivity compared to the high oil flow rates. The variation of the averaged GOR versus oil flow rate of the desalination unit, under the tested conditions, is illustrated in Figure 13. The GOR was increasing at a higher oil flow rate due to the increase in the supply of the thermal energy from the solar concentrator. Additionally, the GOR decreases at a higher circulating air flow rate due to the enhanced cooling effect on the process. Moreover, at oil circulations of 0.984 kg/s and lower, the unit GOR and productivity was significantly reduced, and the process output was adversely affected.

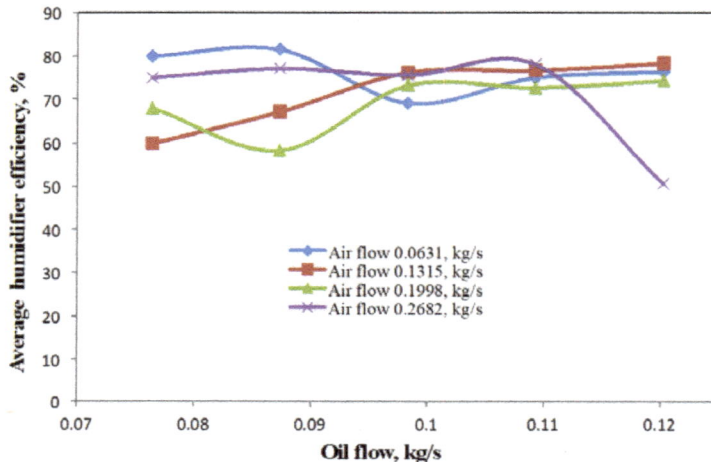

Figure 13. HD average input energy at different air and oil flow rates.

The circulation rates of the oil and air are shown to have higher GORs at an oil flow rate of 0.1202 kg/s, and the process GOR is higher at the lowest circulating air flow rate of 0.0631 kg/s compared with other tested conditions.

4.4. Humidifier Efficiency and Thermal Efficiency

The humidifier efficiency is linked to the design, packing type and air-to-water ratio. All these parameters are reflected on the mass transfer coefficient and therefore shown in the performance efficiency of the humidifier. The variation of the humidifier efficiency at different air to water flow ratios and different heat input conditions, represented by oil flow rates, are shown in Figure 14. The amount of vapor transferred is above 60%, compared to the ideal saturated condition, in order to humidify the air. The thermal efficiency of the humidifier, where the performance appears to be lower when the oil flow rate increases to 0.0983 kg/s and above, is shown in Figure 15. The humidifier has the efficiency of 78% and above to deliver the energy from the heat input to the coldest part, which is the humid air in this condition.

The humidifier average energy-based effectiveness for all cases is shown in Figure 16. At higher circulating air flow rates, the energy-based effectiveness is shown to be decreased. Air flow rate has an influence on the cooling of the system and reduces the actual heat transfer rate. Therefore, a further energy supply is required. For all cases, increasing the oil flow rate has an insignificance effect of energy-based effectiveness reduction compared to the increase to the circulating air flow rate. At the highest test air flow of 0.2682 kg/s, the average energy-based effectiveness is shown to be from 0.4 to 0.476.

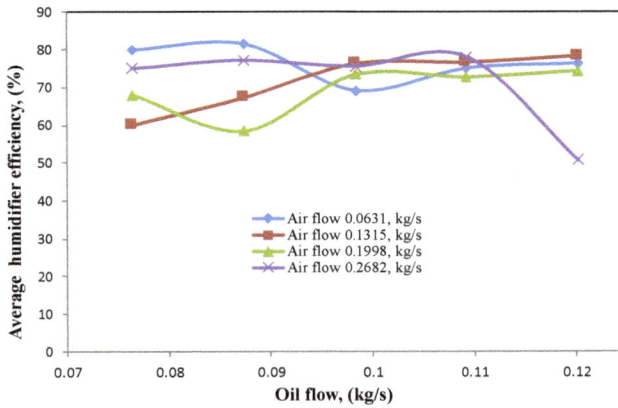

Figure 14. Average humidifier efficiency at different air and oil flow rates.

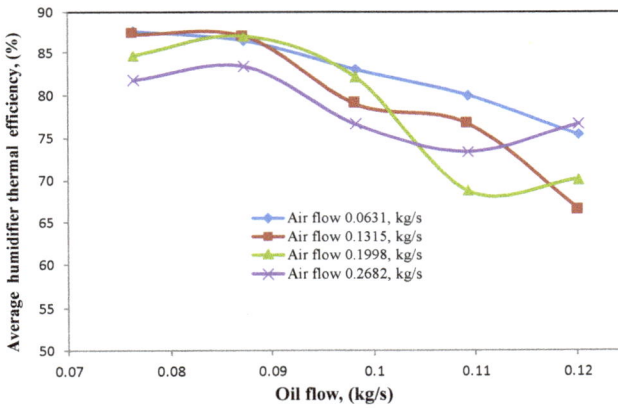

Figure 15. Average humidifier thermal efficiency at different air and oil flow rates.

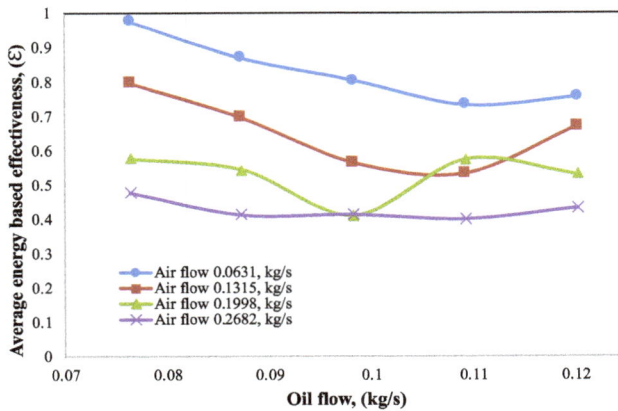

Figure 16. Average energy-based effectiveness ε at different air and oil flow rates.

4.5. Enthalpy Difference

The enthalpy of the humidifier outlet is the combination of the enthalpy of the dry air and the enthalpy of the entrapped water vapor within the stream. The outlet enthalpy is measured relative to the inlet air enthalpy in the humidifier. The calculated enthalpies are related to the specific and latent heats at the operating temperature and the amount of vapor in the streams, which have been computed by Equation (42). The enthalpy difference of the humidifier, at different operating air flow rates and different circulating oil flow rates, is illustrated in Figure 17.

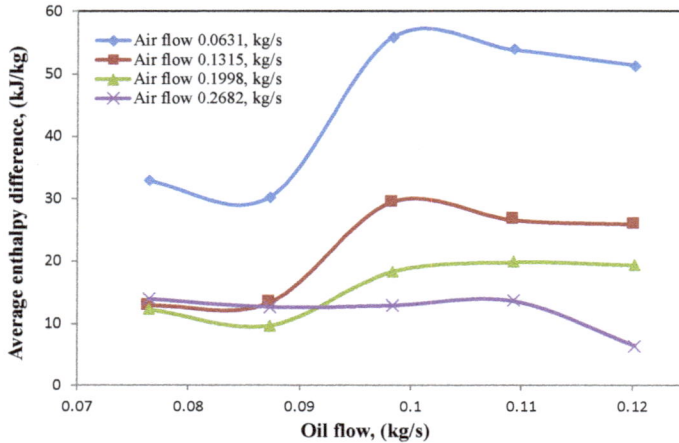

Figure 17. Average enthalpy difference at different air and oil flow rates.

At lower air flow rates, the enthalpy difference of the system is shown to be greater, and the oil circulation rate of 0.0983 kg/s is shown to increase the enthalpy difference for most of the air streams.

4.6. Humidifier Mass Transfer Coefficient

The average mass transfer coefficient of the experiment is plotted against the oil flow rate and is shown in Figure 18. The effect of both heating and air/water circulation on the coefficient is shown in this figure. The air–water flow rate significantly affects the value of the mass transfer coefficient during all experimental runs. Increasing the air flow rate reveals an increase of the mass transfer coefficient up to the air flow of 0.2682 kg/s, at which point the coefficient declined due to concentration and heat potential complications. The process is known to have a mass transfer due to the concentration difference on the airside of the interface and latent heat potentials. The mass transfer coefficient is calculated from Equation (26), in which the process driving force is expressed in terms of the enthalpy difference, where the enthalpy is a thermodynamic property calculated using Equation (42). Hence, the driving force is the function of both temperature and humidity. The variation of the circulating oil flow changes the heating input to the humidifier, which is reflected in the mass transfer behavior. The air-to-water ratio affects the humidity in the system and dominantly controls the mass transfer coefficient values, where the temperature influences the system efficiency when further heat is introduced. At air flow rates of 0.0632 and 0.2682 kg/s, the system could not gain the required heating due to an insufficient air flow, low air flow rate or excessive air flow, which led to a cooling effect at higher air flow rates.

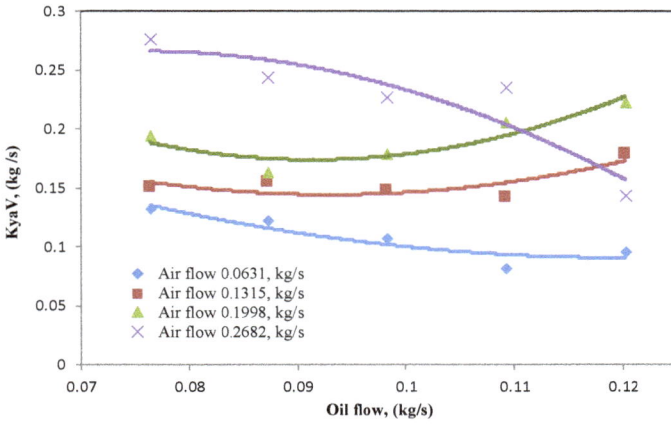

Figure 18. Average HD mass transfer coefficient at different air and oil flow rates.

4.7. Convective Mass Transfer Coefficient

The convective heat to mass transfer coefficient plot, where the coefficient increases as the air flow rate increases, is shown in Figure 19. At lower air flow rates, this ratio decreases when the oil flow rate reaches 0.0983 kg/s, at which point maximum productivity is obtained. This indicates the reduction in the thermal loss and the usefulness of the heat for enhancing the humidifier-driving force through the mass transfer controlled by the humidity difference. At a lower oil flow rate, the process either loses heat or has an insufficient heat input due to higher air flow rates. The air flow rate at 0.2682 shows a higher ratio trend, which can be further improved by using a higher-grade insulation material to enhance the process productivity at higher air flow rates and avoid heat loss. For the present system, the productivity was shown to have declined when the air increased to this limit; hence, optimization is considered a practical case, which could be implemented in order to obtain a tradeoff between the cost, productivity, and size for a good performance. The best unit performance was obtained at the lowest tested air flow and the point at which the oil flow reached 0.0983 kg/s, where the lower loss was obtained per unit volume.

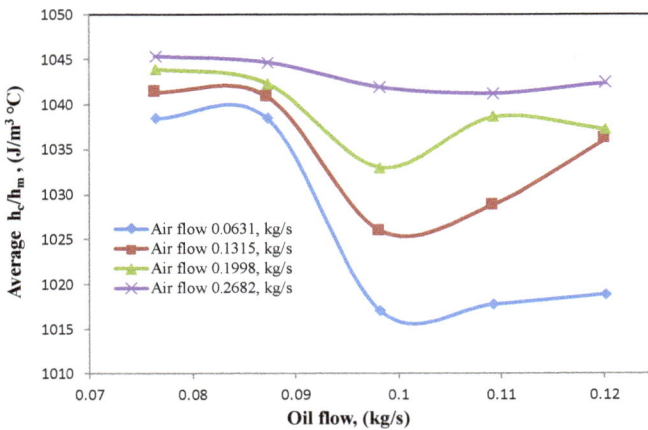

Figure 19. Average HD convective heat and mass transfer coefficient ratio at different air and oil flow rates.

4.8. Optimum Condition

The results of the experiment are analyzed in terms of productivity, production cost and GOR in order to find the best operating conditions that meet the process economics with a high yield. The behavior of the process at different circulating oil and air flow rates, obtaining a tradeoff between the maximum productivity, production cost, and the GOR, is illustrated in Figure 20. Under all operating conditions, the process reveals a high productivity and GOR, with the lowest operating cost, at a circulating oil flow of 0.0983 kg/s. The maximum productivity obtained is 24.31 kg/day at a circulating air flow rate of 0.0631 kg/s, and the GOR is found to be 0.51. This condition has the minimum operating cost, where the production cost attained is 12.54 US\$/m^3, which is 0.0125 US\$/liter. Since the process used solar energy for heating, the heating energy cost would not affect the unit production cost, although the capital investment would. The process economics can be improved by the reduction of the system capital investment, such as automation and controls, since the system works at low pressures and temperatures, or by increasing the productivity by enhancing the power utilization using an auxiliary heating system operated by solar photovoltaic (PV) panels. The general output of the process reduces as the circulating air flow rate increases. At the highest air flow rate, the economics of the process is found to be discouraging, with a production cost of 38.92 US\$/m^3 under the optimal circulating oil condition. Similarly, the GOR of the process has a tendency toward further reduction when the circulating air flow rate is increased. There is also uncertainty in the operating variables due to the variation during different seasons throughout the year, which could affect the overall system results, but the process is expected to behave in a manner similar to the results obtained. The more expensive design may allow for a higher circulating oil temperature of the parabolic concentrator, which requires a larger heat exchanger surface area or humidifier air and water contact area; hence, the benefits could require a higher system capital cost. The maximum output of the experiment, compared with the previous work in the field of solar desalination using the HDH method, is presented in Table 4. Despite the design and capacities of the processes in other research work, the process design in the present study has economic improvements, which enhance the commercialization potential of the process, as demonstrated in Table 4.

Table 4. Comparison between the current result and previous work.

Reference	Description	Productivity, L/day	Cost, US\$
Sharshir et al. [46]	Hybrid solar HDH and solar stills with evacuated solar water heater.	66.3	0.034
Zhani et al. [47]	Flat plate air solar collector and flat-plate water solar collectors with HDH unit.	20	0.093
Hamed et al. [23]	Evacuated tube water solar collector with HDH unit.	22	0.0578
Deniz et al. [48]	Solar air heater and solar water heater with HDH unit.	10.87	0.0981
Zubair et al. [49]	Solar evacuated tubes with HDH unit.	46.59	0.036
Behnam et al. [50]	Solar desalination system equipped with an air bubble column humidifier, evacuated tube collectors and thermosyphon heat pipes.	6.275	0.028
Current work	Solar PTC with HDH unit.	24.31	0.0125

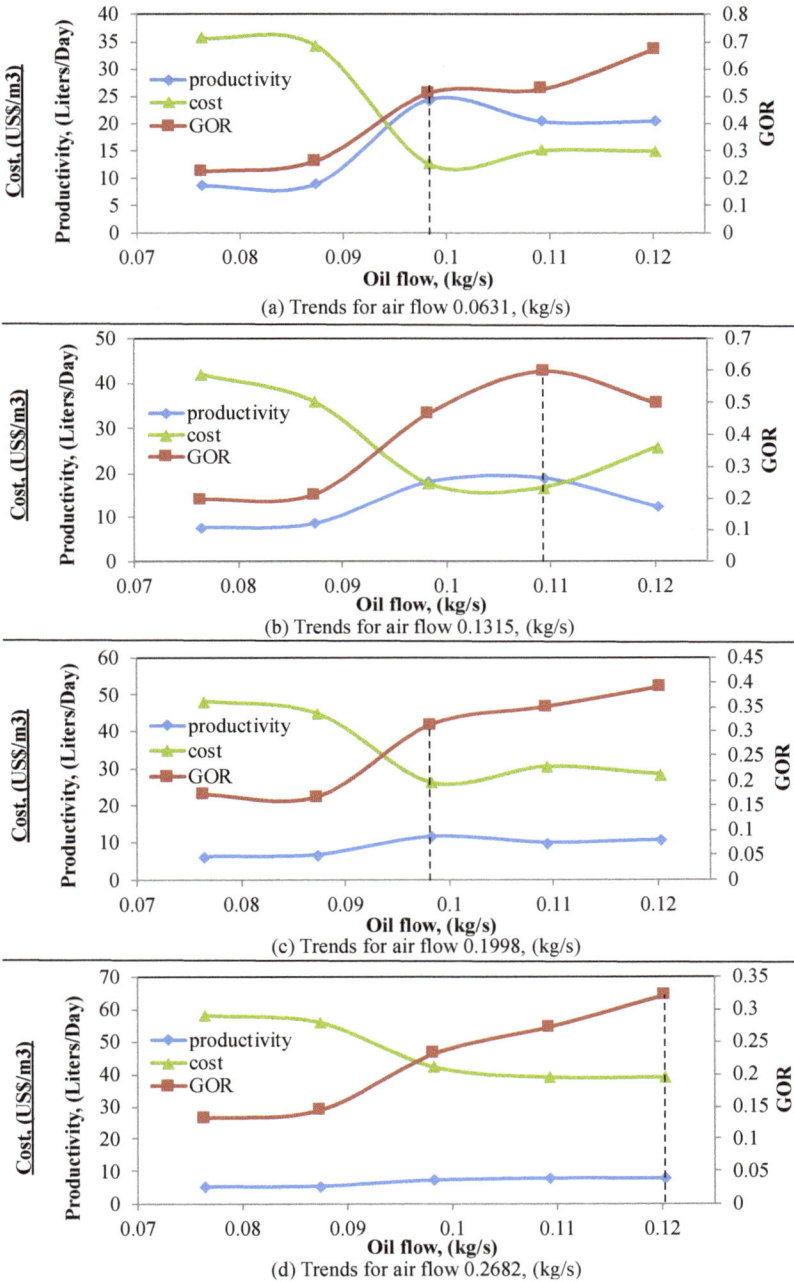

(a) Trends for air flow 0.0631, (kg/s)

(b) Trends for air flow 0.1315, (kg/s)

(c) Trends for air flow 0.1998, (kg/s)

(d) Trends for air flow 0.2682, (kg/s)

Figure 20. System optimization at different air and oil flow rates.

4.9. Collector Efficiency

The behavior of the averaged efficiency of the PTC under the operating conditions is illustrated in Figure 21. A decrease of the instantaneous efficiency of the PTC, following the increase of the circulating oil flow rate, is shown in this figure. The low efficiencies at high circulating oil flow rates, attributed to the shorter residence time of oil in the absorber tubes, reduce the heat gain of the PTC. The circulating water temperature lines of the humidifier, shown in Figure 19, illustrate the condition of the humidification process within the operating oil flow, where the lower oil flow revealed a lower productivity and high PTC efficiency. Moreover, the temperatures of the circulating water in the humidifier were found to be higher at the oil flow rate of 0.0983 kg/s than with other settings, which shows the optimum oil flow rate for the heat exchange in the humidifier. The average PTC efficiency at the lowest oil flow rate is 27.8% and is reduced by 49.5% at the maximum flow rate.

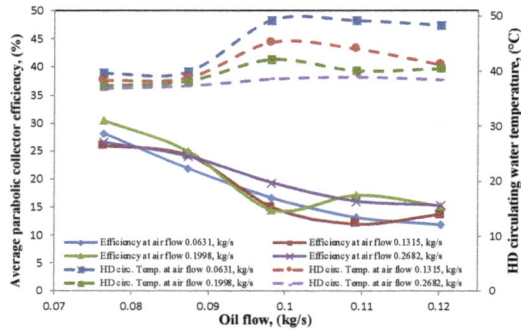

Figure 21. Average collector efficiency at different air and oil flow rates.

In the designed system, the ratio of the heat gained to the amount of beam radiation concentrated in the absorber tube is revealed to be higher, especially from the middle to the end of the solar day due to closed system circulation. Moreover, for all of the cases in which the incidence angle increased throughout the day, the losses due to the end effect decreased; hence, the PTC efficiency increased. When the beam radiation reduced at the end of the day, the system still can deliver the energy to the humidifier at a higher rate due to the stored heat in the oil expansion tank and the circulating air and water of the humidifier. The behavior of the circulating oil inlet and outlet temperatures in the parabolic concentrators is shown in Figure 22. Generally, at higher oil flow rates, the system heat storage capabilities increased due to multiple circulation procedures. This positively affects the value of the parabolic concentrator efficiency; hence, the minimum beam radiation utilization is obtained.

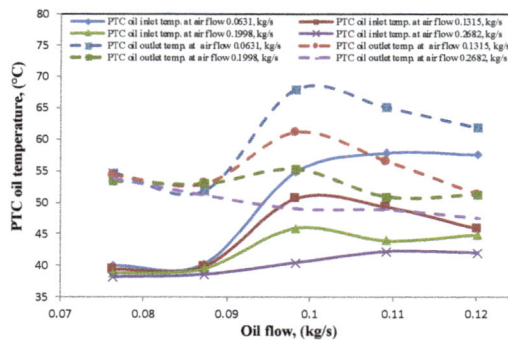

Figure 22. Average PTC oil temperature at different air and oil flow rates.

The heat removal factor F_R is known to be the measure of the thermal resistance by the absorbed radiation in the circulating oil [27]. The heat removal factor changes at different oil and air flow rates, as shown in Figure 23. At very high oil flow rates, it possible that the oil outlet temperature becomes closer to that of the inlet; therefore, the factor is reduced. From the figure, at the highest oil flow rate of 0.12 kg/s, the oil temperature difference decreases and reduces the heat removal factor. In all the tested cases, the oil flow rate of 0.0983 kg/s was shown to have high system productivity and maintained oil temperature difference, and the higher air flow rates revealed higher values of the heat removal factor.

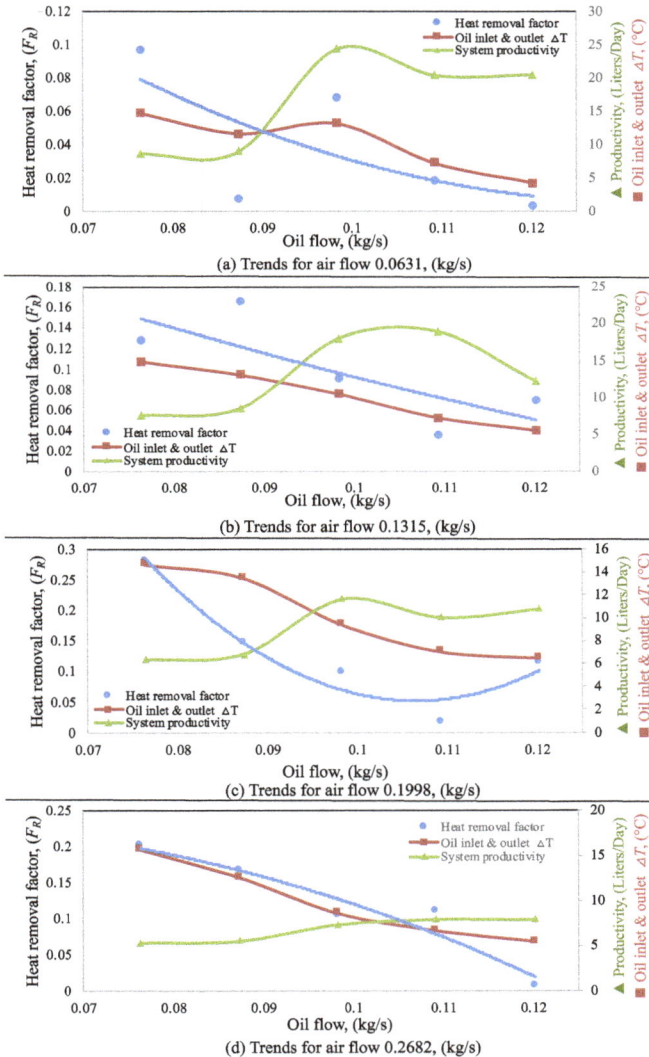

Figure 23. Heat removal factor F_R of the PTCs with the change of system productivity and oil ΔT at different air and oil flow rates.

4.10. Absorber Tube Overall Heat Transfer Coefficient

The overall heat transfer coefficient, including the absorber tube wall, calculated using the heat removal factor, the heat removal coefficient and the convective heat transfer coefficient inside the absorber tube, is obtained from the recorded experimental data. The behavior of the overall heat transfer coefficient, at different circulating oil flow rates and humidifier air flow rates, is illustrated in Figure 24. The circulating oil inside the tube is found to be laminar in all the tested flows, and the corresponding Nusselt number was 4.364. The overall heat transfer coefficient depends on the conduction through the tube wall, the convection inside the tube and the heat removal loss coefficient, and the ambient conditions, such as the wind speed and temperature, influence its value. The convection coefficient for the laminar flow has less significance for the value of the overall heat transfer coefficient, and the wall conduction coefficient does not have an impact on changing the overall heat transfer coefficient. Therefore, the calculated heat removal coefficient profoundly influences the process affected by the ambient conditions. Increasing the circulating oil flow rate will increase the tendency toward a higher heat loss; hence, the overall heat transfer coefficient will increase. At a lower air flow rate, such as 0.0631 kg/s, the absorber tube will gain more heat at a lower oil residence time inside the absorber tube; hence, the system will have a tendency to increase the heat removal coefficient. The heat removal value could be decreased using a glass vacuum tube instead of a PTC glass cover.

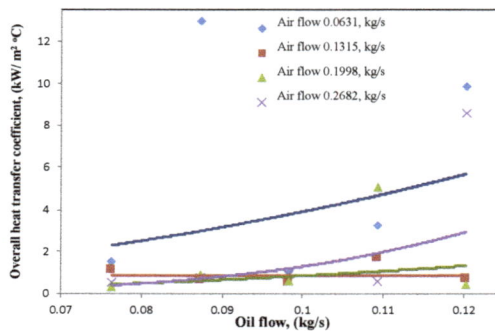

Figure 24. Average absorber tube overall heat transfer coefficient.

5. Conclusions and Recommendations

5.1. Conclusions

This work has integrated modeling, techno-economic analysis, design, and experimental work for the assessment and optimization of a solar-driven humidification and dehumidification desalination unit for arid areas. The system is characterized by high efficiency, relatively simple operation and low maintenance. The GOR of the process, considering the primary purpose of the system construction, would have less priority compared to the other studied factors. Solar energy, concentrated by a PTC system, and an associated HDH desalination unit have been studied in order to evaluate the system productivity, efficiency, and economic feasibility. The design, construction, and testing have been implemented, and the following results have been obtained for a case study in Saudi Arabia:

- The HDH system productivity is significantly influenced by the circulating air-to-water flow ratio, and the highest system productivity obtained is 24.31 kg/day. The best productivity curve for all the tested air flow rates is obtained at a 0.0983 kg/s circulating oil flow rate. The system showed a drastic reduction in productivity of 70.1% when the air flow rate increased to 0.2683 kg/s.
- Increasing the humidifier circulating water temperature positively affects the humidifier productivity. This has been achieved by modulating the heat input to the humidifier by altering

the oil circulation through the PTCs. The maximum humidifier circulating water temperature with the highest productivity is 49.23 °C. The circulating water temperature is reduced by 21.7% when the highest circulating air flow rate of 0.2683 kg/s is achieved.

- Increasing the circulating oil flow rate positively influences the system's productivity until the point when this starts to decline; hence, a tradeoff between either a larger size humidifier heat exchanger or larger PTC surface area is required to enhance productivity.

- The energy consumption of the process under the best conditions varies between 2.39 and 3.07 kWh, and the energy delivered to the humidifier is decreased, when further circulating oil flow is increased, and reached a 39.3% reduction at the highest oil flow rate.

- The humidifier efficiency ranges between 69.1% and 75.6%, and the thermal efficiencies are in the range of 76.6% and 82.94% at a circulating oil flow rate of 0.0983 kg/s.

- The cost analysis of the system indicates a reduction of the production cost under the highest productivity condition. A tradeoff between the cost, productivity, and GOR is necessary. The production cost is 12.54 US\$/m^3 under a condition where the highest GOR of 0.51 is obtained, and the maximum productivity is achieved.

- The efficiency of the PTCs is calculated, and the results showed a significant increase in the performance at a low circulating oil flow rate. The oil circulation procedure in a closed loop enhances the system efficiency by reducing the losses and enhancing the energy storage in the oil expansion tank.

Finally, the PTC coupled with the HDH desalination system reveals a promising desalination technique in arid areas, where no commercial desalination plants are available for a small population. The system outcome can be improved by the further development of a hybrid system of PV, PTC, and auxiliary heaters in combination with an HDH desalination unit.

5.2. Recommendations for Future Work

The solar-driven HDH desalination process would be more effective, with a higher performance, cost-efficient and positive environmental impact in the desalination technology, if the following recommendations were implemented.

- The solar-driven HDH system can be coupled with other thermal desalination technologies to increase the desalination process yield. The dual-purpose plants for power and desalination can be significantly improved. The optimization of the dual-purpose system can benefit from and be enhanced using a shortcut method to address the need for conceptual design studies [51]. Moreover, decoupling the mass balances and the heat balances and heat-transfer sizing equations using the linear programming (LP) optimization formulation simplifies the computations and significantly reduces the model size and complexity [52]. A hybrid design can include both thermal and reverse osmosis plants [53]; therefore, economic optimization can be carried out in the overall system to identify the process configuration tradeoffs in various scenarios. Mixed-integer nonlinear programming (MINLP) is used for 20 RO membrane modules, and 58% of savings in freshwater use is achieved as compared to the existing base case operations [54]. The solution to this program provides the optimal arrangement, types, and sizes of the reverse-osmosis units and the booster pumps [55]. Such an application showed an improvement in the optimization, determining the optimal mix of solar energy, thermal storage, and fossil fuel to attain the maximum annual profit of the shale gas production system studied [56], which can be applied to the solar-driven HDH system studied.

- Heat integration can be performed to reduce the operating cost by coupling the HDH desalination system to industrial facilities. This approach has been implemented along with the optimization of multi-effect distillation (MED) and membrane distillation (MD) and revealed a successful cost reduction in different scenarios [57]. Moreover, coupling multi-stage flash (MSF) with MED desalination processes was shown to improve the system output due to the use of the waste heat recovery [58]. Therefore, the HDH desalination can be integrated with solar collectors, absorption refrigeration cycles, and organic Rankine cycles.

- The existing desalination systems can consider the HDH as a modular desalination technology in the retrofitting option, which increases the capacity of the output [59].
- The economic factors are the most important factors relating to project improvement; hence, the low cost-integrated solar HDH desalination system is an option that incorporates sustainability in the design and selection of projects. The metric approach can provide an evaluation of project profitability and promote sustainability related to the capital cost [60].

Author Contributions: Conceptualization, F.A. and M.A.; Methodology, F.A.; Software, F.A.; Validation, F.A., A.M. and M.A.; Formal Analysis, M.A.; Investigation, M.A.; Resources, F.A.; Data Curation, M.A.; Writing-Original Draft Preparation, M.A.; Writing-Review & Editing, A.M.; Visualization, F.A.; Supervision, F.A.; Project Administration, A.A.; Funding Acquisition, F.A.

Funding: This project was funded by the National Plan for Science, Technology and Innovation (MAARIFAH)—King Abdulaziz City for Science and Technology—the Kingdom of Saudi Arabia—award number (11-ENE2004-03).

Acknowledgments: The authors also acknowledge, with thanks, the Science and Technology Unit, King Abdulaziz University for their technical support. The authors acknowledge the great support of Abdulsalam Alghamdy, head of the King Salman Energy chair, in accommodating the project installation in their premises.

Conflicts of Interest: The authors declare no conflict of interest.

Nomenclature

A	Surface area, m^2
A_a	Total aperture area, m^2
A_e	Lost area due to end effect, m^2
A_f	Geometric ratio, m^2
A_l	Lost area, m^2
C	Collector concentration ratio
C_A	Annual capital cost, US$
C_M	Maintenance cost, US$
C_{OP}	Operating cost, US$
C_r	Heat capacity ratio
C_{TOP}	Total operating cost, US$
D_{AB}	Mass diffusion coefficient of air in water, m^2/s
D_i	Inside diameter, m
D_o	Outer diameter, m
F_R	PTC heat removal factor
G_{bn}	Normal beam radiation, W/m^2
H	Enthalpy, kJ/kg
I	Solar irradiance, W/m^2
K_y	Gas phase mass transfer coefficient of humidifier, Kg/m^2 s
L	Latitude angel, degrees
N	Number of the day in the Gregorian year
Nu	Nusselt number
P	Capital cost, US$
Pr	Prandtl number
P_T	Total pressure, mmHg
P_v	Vapor pressures, mmHg
Q_{in}	Heat transfer rate, W
Re	Reynolds number
Sc	Schmidt number
T_a	Air temperature, °C
T_{cw}	Cooling water temperature, °C
T_{MU}	Make water temperature, °C

T_{oil}	Oil temperature, °C
T_w	Water temperature, °C
U	Overall heat transfer coefficient, W/m^2 °C
U_c	Overall heat transfer coefficient of the absorber tube, W/m^2 °C
U_L	Heat removal coefficient of the absorber tube, W/m^2 °C
V	Volume of the humidifier, m^3
W	Aperture width, m
a	Interface area, m^{-1}
c_a	Specific heat of humid air, kJ/kg °C
c_w	Specific heat of water, kJ/kg °C
f_c	Enthalpy correction factor, kJ/kg
h	Hour angle, degrees
h_{ab}	Oil convective heat transfer coefficient, kJ/s m^2 °C
h_c	Convective heat transfer coefficient, kJ/s m^2 °C
h_{fg}	Latent heat of evaporation of water at ambient conditions, kJ/kg
h_m	Convective mass transfer coefficient, m/s
h_p	Hight of parabola, m
i	Interest rate, %
k	Thermal conductivity, W/m °C
k_{ab}	Absorber tube thermal conductivity, kJ/s m^2 °C
l	Trough length, m
m_e	Air flow rate, Kg/s
m_{cw}	Cooling water flow rate, Kg/s
m_w	Water flow rate, Kg/s
m_d	Product water flow rate, Kg/s
m_{MU}	Make up water flow rate, Kg/s
m_{oil}	Oil flow rate, Kg/s
n	Number of years
t_s	Solar time, hour

Indices

a	Air
amb	Ambient condition
as	Saturated air
avg	Average value
DH	Dehumidifier
HD	Humidifier
lm	Log mean
m	Mean value
oil	Oil side
s	Saturated
v	vapor
w	water

Greek

α	Altitude angel, degrees
β	Slop of the PTC surface, degrees
γ	Intercept factor
δ	Declination angle, degrees
ε	Heat exchanger effectiveness
ζ	Reflectance of the reflector
η	PTC efficiency, %
η_H	Humidifier efficiency, %
$\eta_{T_{HD}}$	Humidifier thermal efficiency, %
η_o	Optical efficiency, %
θ	The incident angle, degrees
θ_z	Zenith angle, degrees

λ_0	Latent heat of evaporation of water at base temperature, kJ/kg
μ	Dynamic viscosity, kg/m s
ρ	Density, kg/m^3
τ	Transmittance of the glass cover
ϕ	Azimuth angle, degrees
ϕ_s	Surface azimuth angle, degrees
ψ	Absorptance of the receiver
ω	Absolute humidity, kg$_{water}$/kg$_{air}$

Appendix A

The standard uncertainties for the measured data are obtained based on the type A evaluation procedure, described in the JCGM 100:2008 [61]. The standard uncertainties, obtained from the statistical analysis of the experimental data, are shown in Figure A1. The calculation is based on the set of 14 values, measured repeatedly at noon; hence, the standard deviation has 13 degrees of freedom for each measurement. The maximum standard uncertainty for the measured data is found to be 0.94.

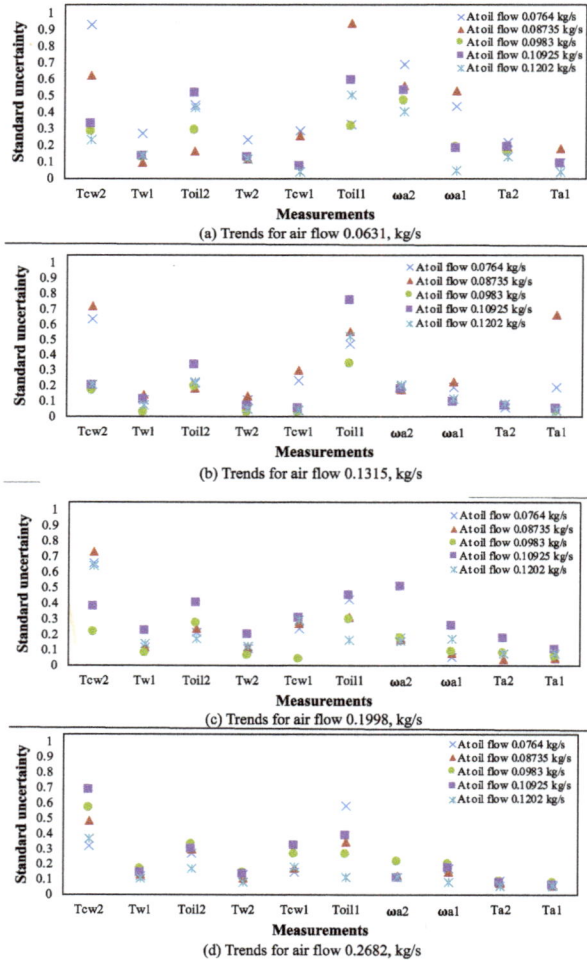

(a) Trends for air flow 0.0631, kg/s

(b) Trends for air flow 0.1315, kg/s

(c) Trends for air flow 0.1998, kg/s

(d) Trends for air flow 0.2682, kg/s

Figure A1. The standard uncertainty for the experimental measurements.

References

1. Al-Rashed, M.F.; Sherif, M.M. Water resources in the GCC countries: An overview. *Water Resour. Manag.* **2000**, *14*, 59–75. [CrossRef]
2. Almasoud, A.H.; Gandayh, H.M. Future of solar energy in Saudi Arabia. *J. King Saud Univ. Eng. Sci.* **2015**, *27*, 153–157. [CrossRef]
3. Zell, E.; Gasim, S.; Wilcox, S.; Katamoura, S.; Stoffel, S.; Shibli, H.; Engel-Cox, J.; Al Subie, M. Assessment of solar radiation resources in Saudi Arabia. *Sol. Energy* **2015**, *119*, 422–438. [CrossRef]
4. Pazheri, F.R. Solar power potential in Saudi Arabia. *Int. J. Eng. Res. Appl.* **2014**, *4*, 171–174.
5. Solar GIS, Solar Resource Maps of Saudi Arabia. Available online: https://solargis.com/maps-and-gis-data/download/saudi-arabia (accessed on 5 May 2018).
6. Tzivanidis, C.; Bellos, E.; Korres, D.; Antonopoulos, K.A.; Mitsopoulos, G. Thermal and optical efficiency investigation of a parabolic trough collector. *Case Stud. Therm. Eng.* **2015**, *6*, 226–237. [CrossRef]
7. Yassen, T.A. Experimental and theoretical study of a parabolic trough solar collector. *Anbar J. Eng. Sci.* **2012**, *5*, 109–125.
8. Murtuza, S.A.; Byregowda, H.V.; Imran, M. Experimental and simulation studies of parabolic trough collector design for obtaining solar energy. *Resour. Eff. Technol.* **2017**, *3*, 414–421. [CrossRef]
9. Jebasingh, V.K.; Joselin, G.M. A review of solar parabolic trough collector. *Renew. Sustain. Energy Rev.* **2016**, *54*, 1085–1091. [CrossRef]
10. Arun, C.A.; Sreekumar, P.C. Modeling and performance evaluation of parabolic trough solar collector desalination system. *Mater. Today Proc.* **2018**, *5*, 780–788. [CrossRef]
11. Elashmawy, M. An experimental investigation of a parabolic concentrator solar tracking system integrated with a tubular solar still. *Desalination* **2017**, *411*, 1–8. [CrossRef]
12. Zohreh, R.; Hatamipour, M.S.; Ghalavand, Y. Solar assisted modified variable pressure humidification-dehumidification desalination system. *Energy Convers. Manag.* **2018**, *162*, 321–330.
13. Mohamed, A.M.I.; El-Minshawy, N.A. Theoretical investigation of solar humidification–dehumidification desalination system using parabolic trough concentrators. *Energy Convers. Manag.* **2011**, *52.10*, 3112–3119. [CrossRef]
14. Fathy, M.; Hamdy, H.; Salem, M. Experimental study on the effect of coupling parabolic trough collector with double slope solar still on its performance. *Sol. Energy* **2018**, *163*, 54–61. [CrossRef]
15. Kabeel, A.E.; Hamed, M.H.; Omara, Z.M.; Sharshir, S.W. Water desalination using a humidification-dehumidification technique—A detailed review. *Nat. Resour.* **2013**, *4*, 286–305. [CrossRef]
16. Srithar, K.; Rajaseenivasan, T. Performance analysis on a solar bubble column humidification dehumidification desalination system. *Process Saf. Environ. Prot.* **2017**, *105*, 41–50. [CrossRef]
17. Zubair, S.M.; Antar, M.A.; Elmutasim, S.M.; Lawa, D.U. Performance evaluation of humidification-dehumidification (HDH) desalination systems with and without heat recovery options: An experimental and theoretical investigation. *Desalination* **2018**, *436*, 161–175. [CrossRef]
18. Hamed, M.H.; Kabeel, A.E.; Omara, Z.M.; Sharshir, S.W. Mathematical and experimental investigation of a solar humidification–dehumidification desalination unit. *Desalination* **2015**, *358*, 9–17. [CrossRef]
19. Ahmed, H.A.; Ismail, I.M.; Saleh, W.F.; Ahmed, M. Experimental investigation of humidification-dehumidification desalination system with corrugated packing in the humidifier. *Desalination* **2017**, *410*, 19–29. [CrossRef]
20. Ghaffour, N.; Bundschuh, J.; Mahmoudi, H.; Goosen, M.F. Renewable energy-driven desalination technologies: A comprehensive review on challenges and potential applications of integrated systems. *Desalination* **2015**, *356*, 94–114. [CrossRef]
21. Li, C.; Goswami, Y.; Stefanakos, E. Solar assisted sea water desalination: A review. *Renew. Sustain. Energy Rev.* **2013**, *19*, 136–163. [CrossRef]
22. Boyes, W. *Instrumentation Reference Book*, 3rd ed.; Elsevier Butterworth-Heinemann: Burlington, MA, USA, 2003; p. 272.
23. Kalogirou, S.A. *Solar Energy Engineering: Processes and Systems*, 2nd ed.; Academic Press: Waltham, MA, USA, 2013.
24. Rabl, A. *Active Solar Collectors and Their Applications*; Oxford University Press: New York, NY, USA, 1985.

25. Camacho, E.F.; Berenguel Soria, M.; Rubio, F.R.; Martínez, D. *Control of Solar Energy Systems*; Springer Science & Business Media: New York, NY, USA, 2012; pp. 25–47.
26. Wang, S.K. *Handbook of Air Conditioning and Refrigeration*, 2nd ed.; McGraw-Hill: New York, NY, USA, 2001.
27. Coccia, G.; Di Nicola, G.; Hidalgo, A. *Parabolic trough Collector Prototypes for Low-Temperature Process Heat*; Springer International Publishing AG Switzerland: Basel, Switzerland, 2016.
28. Goswami, D.Y.; Kreith, F.; Kreider, J.F. *Principles of Solar Engineering*, 2nd ed.; CRC Press (Taylor and Francis Group): Philadelphia, PA, USA, 2000.
29. Kreith, F.; Jan, F.K. *Principles of Solar Engineering*; Hemisphere Pub. Corp.: Washington, WN, USA, 1978.
30. Duffie, J.A.; Beckman, W.A. *Solar Engineering of Thermal Processes*; John Wiley & Sons: Hoboken, NJ, USA, 2013.
31. Abdel-Hady, F.; Shakil, S.; Hamed, M.; Alzahrani, A.; Mazher, A. Simulation and Manufacturing of an Integrated Composite Material Parabolic Trough Solar Collector. *Int. J. Eng. Technol.* **2016**, *8*, 2333–2345. [CrossRef]
32. Incropera, F.P.; Dewitt, D.P.; Bergman, T.L.; Lavine, A.S. *Fundamentals of Heat and Mass Transfer*, 6th ed.; John Wiley & Sons: Danvers, MA, USA, 2007.
33. Bejan, A.; Kraus, A.D. *Heat Transfer handbook*; John Wiley & Sons: Hoboken, NJ, USA, 2003; Volume 1.
34. Kröger, D.G. *Air-Cooled Heat Exchangers and Cooling Towers. Thermal-Flow Performance Evaluation and Design*; PennWell Corporation: Tulsa, OK, USA, 2004; Volume 1.
35. Jaber, H.; Webb, B.L. Design of Cooling Towers by the Effectiveness-*NTU* Method. *ASME J. Heat Transf.* **1989**, *111*, 837–843. [CrossRef]
36. Narayan, G.P.; Mistry, K.H.; Sharqawy, M.H.; Zubair, S.M.; Lienhard, J.H. Energy effectiveness of simultaneous heat and mass exchange devices. *Front. Heat Mass Transf.* **2010**, *1*, 023001. [CrossRef]
37. Kreith, F. *The CRC Handbook of Thermal Engineering*; CRC Press LLC: Boca Raton, FL, USA, 2000.
38. Çengel, Y.A.; Boles, M.A. *Thermodynamics: An Engineering Approach*, 5th ed.; McGraw-Hill: New York, NY, USA, 2006.
39. Sinnott, R.K. *Coulson & Richardson's Chemical Engineering, Chemical Engineering Design*, 4th ed.; Elsevier Butterworth-Heinemann: Oxford, UK, 2005; Volume 6.
40. Kabeel, A.E.; El-Said, E.M.S. Applicability of flashing desalination technique for small scale needs using a novel integrated system coupled with nanofluid-based solar collector. *Desalination* **2014**, *333*, 10–22. [CrossRef]
41. Towler, G.; Sinnott, R. *Chemical Engineering Design Principles, Practice and Economics of Plant and Process Design*; Elsevier Butterworth-Heinemann: Burlington, MA, USA, 2008.
42. El-Halwagi, M.M. Introduction to Sustainability, Sustainable Design, and Process Integration. In *Sustainable Design through Process Integration: Fundamentals and Applications to Industrial Pollution Prevention, Resource Conservation, and Profitability Enhancement*, 2nd ed.; IChemE, Elsevier: New York, NY, USA, 2017.
43. El-Dessouky, H.T.; Ettouney, H. *Fundamentals of Salt Water Desalination*; Elsevier Science B.V.: Amsterdam, The Netherlands, 2002.
44. Khisty, J.C.; Mohammadi, J.; Amekudzi, A.A. *Systems Engineering with Economics, Probability and Statistics*, 2nd ed.; J. Ross Publishing: Lauderdale, FL, USA, 2012.
45. Saudi Electricity Company Official Web Site. Available online: https://www.se.com.sa/en-us/Customers/Pages/TariffRates.aspx (accessed on 12 March 2018).
46. Sharshir, S.W.; Peng, G.; Yang, N.; Eltawil, M.A.; Ali, M.K.A.; Kabeel, A.E. A hybrid desalination system using humidification-dehumidification and solar stills integrated with evacuated solar water heater. *Energy Convers. Manag.* **2016**, *124*, 287–296. [CrossRef]
47. Zhani, K.; Bacha, H.B. Experimental investigation of a new solar desalination prototype using the humidification dehumidification principle. *Renew. Energy* **2010**, *35*, 2610–2617. [CrossRef]
48. Deniz, E.; Çınar, S. Energy, exergy, economic and environmental (4E) analysis of a solar desalination system with humidification-dehumidification. *Energy Convers. Manag.* **2016**, *126*, 12–19. [CrossRef]
49. Zubair, M.I.; Al-Sulaiman, F.A.; Antar, M.A.; Al-Dini, S.A.; Ibrahim, N.I. Performance and cost assessment of solar driven humidification dehumidification desalination system. *Energy Convers. Manag.* **2017**, *132*, 28–39. [CrossRef]
50. Behnam, P.; Shafii, M.B. Examination of a solar desalination system equipped with an air bubble column humidifier, evacuated tube collectors and thermosyphon heat pipes. *Desalination* **2016**, *397*, 30–37. [CrossRef]

51. El-Halwagi, M.M. A Shortcut Approach to the Design of Once-Through Multi-Stage Flash Desalination Systems. *Desalin. Water Treat.* **2017**, *62*, 43–56. [CrossRef]

52. Gabriel, K.J.; Noureldin, M.M.B.; Linke, P.; El-Halwagi, M.M. Optimization of Multi-Effect Distillation Process Using a Linear Enthalpy Model. *Desalination* **2015**, *365*, 261–276. [CrossRef]

53. Gabriel, K.; El-Halwagi, M.M.; Linke, P. Optimization Across Water-Energy Nexus for Integrating Heat, Power, and Water for Industrial Processes Coupled with Hybrid Thermal-Membrane Desalination. *Ind. Eng. Chem. Res.* **2016**, *55*, 3442–3466. [CrossRef]

54. Khor, C.S.; Foo, D.C.Y.; El-Halwagi, M.M.; Tan, R.R.; Shah, N. A Superstructure Optimization Approach for Membrane Separation-Based Water Regeneration Network Synthesis with Detailed Nonlinear Mechanistic Reverse Osmosis Model. *Ind. Eng. Chem. Res.* **2011**, *50*, 13444–13456. [CrossRef]

55. El-Halwagi, M.M. Synthesis of Optimal Reverse-Osmosis Networks for Waste Reduction. *AIChE J.* **1992**, *38*, 1185–1198. [CrossRef]

56. Al-Aboosi, F.Y.; El-Halwagi, M.M. An Integrated Approach to Water-Energy Nexus in Shale Gas Production. *Processes* **2018**, *6*, 52. [CrossRef]

57. Bamufleh, H.; Abdelhady, F.; Baaqeel, H.M.; El-Halwagi, M.M. Optimization of Multi-Effect Distillation with Brine Treatment via Membrane Distillation and Process Heat Integration. *Desalination* **2017**, *408*, 110–118. [CrossRef]

58. González-Bravo, R.; Ponce-Ortega, J.M.; El-Halwagi, M.M. Optimal Design of Water Desalination Systems Involving Waste Heat Recovery. *Ind. Eng. Chem. Res.* **2017**, *56*, 1834–1847. [CrossRef]

59. Baaqeel, H.; El-Halwagi, M.M. Optimal Multi-Scale Capacity Planning in Seawater Desalination Systems. *Processes* **2018**, *6*, 68. [CrossRef]

60. El-Halwagi, M.M. A Return on Investment Metric for Incorporating Sustainability in Process Integration and Improvement Projects. *Clean Technol. Environ. Policy* **2017**, *19*, 611–617. [CrossRef]

61. BIPM; IFCC; IUPAC; ISO. *Evaluation of Measurement Data Guide for the Expression of Uncertainty in Measurement*; JCGM 100: Sèvres Cedex, France, 2008.

processes

MDPI

Article

Intensification of Reactive Distillation for TAME Synthesis Based on the Analysis of Multiple Steady-State Conditions

Takehiro Yamaki [1,*][iD], Keigo Matsuda [2,3,*][iD], Duangkamol Na-Ranong [4] and Hideyuki Matsumoto [5]

1   Research Institute for Chemical Process Technology, National Institute of Advanced Industrial Science and Technology (AIST), Central 5, 1-1-1 Higashi, Tsukuba-shi, Ibaraki 305-8565, Japan
2   Department of Chemistry and Chemical Engineering, Graduate School of Science and Engineering, Yamagata University, 4-3-16 Jonan, Yonezawa-shi, Yamagata 992-8510, Japan
3   Renewable Energy Center, National Institute of Advanced Industrial Science and Technology (AIST), 2-2-9, Machiikedai, Koriyama, Fukushima 963-0298, Japan
4   Department of Chemical Engineering, Faculty of Engineering, King Mongkut's Institute of Technology Ladkrabang, Chalongkrung Road, Bangkok 10520, Thailand; dnaranong@hotmail.com
5   Department of Chemical Science and Engineering, School of Materials and Chemical Technology, Tokyo Institute of Technology, 2-12-1 Ookayama, Meguro-ku, Tokyo 152-8552, Japan; hmatsumo@chemeng.titech.ac.jp
*   Correspondence: takehiro-yamaki@aist.go.jp (T.Y.); matsuda@yz.yamagata-u.ac.jp or matsuda-k@aist.go.jp (K.M.); Tel.: +81-29-861-3695 (T.Y.); +81-238-26-3742 (K.M.)

Received: 31 October 2018; Accepted: 20 November 2018; Published: 26 November 2018

Abstract: Our previous study reported that operation in multiple steady states contributes to an improvement in reaction conversion, making it possible to reduce the energy consumption of the reactive distillation process for *tert*-amyl methyl ether (TAME) synthesis. This study clarified the factors responsible for an improvement in the reaction conversion for operation in the multiple steady states of the reactive distillation column used in TAME synthesis. The column profiles for those conditions, in which multiple steady states existed and those in which they did not exist, were compared. The vapor and liquid flow rates with the multiple steady states were larger than those when the multiple steady states did not exist. The effect of the duty of the intermediate condenser, which was introduced at the top of the reactive section, on the liquid flow rate for a reflux ratio of 1 was examined. The amount of TAME production increased from 55.2 to 72.1 kmol/h when the intermediate condenser was operated at 0 to −5 MW. Furthermore, the effect of the intermediate reboiler duty on the reaction performance was evaluated. The results revealed that the liquid and vapor flow rates influenced the reaction and separation performances, respectively.

Keywords: reactive distillation; multiple steady state; steady state simulation; reaction conversion; TAME synthesis

---

1. Introduction

In the chemical industry, high-efficiency reactions and separation processes are needed to reduce greenhouse gas emissions. Reactive separation processes have attracted considerable attention given their high levels of efficiency [1]. Reactive distillation is one such reactive separation technology. Reactive distillation offers advantages such as energy and capital savings, increased reaction conversion, high selectivity, and utilization of the reaction heat [2]. Therefore, reactive distillation processes have been examined for application to esterification and etherification reactions [3–6].

The design and operability of a reactive distillation process become more difficult than those of a conventional distillation or reaction process because of the interactions between the reaction and the separation. Multiple steady states exist in a reactive distillation process because of these interactions. Through experiments, Mohl et al. confirmed the existence of multiple steady states when using a pilot-scale reactive distillation column for *tert*-amyl methyl ether (TAME) synthesis [7]. Some researchers empirically confirmed the existence of multiple steady states in a reactive distillation process. Jacobs and Krishna and Nijhuis et al. reported on the existence of multiple steady states observed in the simulation of a reactive distillation for the synthesis of methyl *tert*-butyl ether (MTBE) using a steady-state equilibrium stage model [8,9]. Baur et al. performed bifurcation analysis for TAME synthesis in a reactive distillation using the reaction kinetics methods for pseudo-homogeneous and heterogeneous models [10]. They reported that multiple steady states can be captured by simulation for two different reaction kinetics methods. Wang et al. analyzed the product purity attained with MTBE synthesis in a reactive distillation column and observed multiple steady states [11]. Ramzan et al. simulated a reactive distillation column for ethyl *tert*-butyl ether synthesis and again detected multiple steady states [12]. Cárdenas-Guerra et al. found multiple steady states in the reactive distillation process for the deep hydrodesulfurization of diesel [2]. Meanwhile, Jairnel-Leal et al. analyzed the operating conditions and parameter sensitivity of multiple steady states in a reactive distillation for TAME and MTBE syntheses [13]. They found that the feed thermal condition has a major influence on the occurrence of multiple steady states.

Our research group considers that the process performance of a reactive distillation can be improved through the multiple steady-state condition, and process design and analysis are performed. The effects of the operating conditions in the multiple steady states on the performance of a reactive distillation column used for TAME synthesis were evaluated [14]. The reaction conversion in the upper branch in the multiple steady states was the highest and increased with the reflux ratio. In addition, the energy-saving performance of reactive distillation consisting of one reactive distillation column and two recovery distillation columns for the multiple steady states was examined [15]. The energy consumption for TAME synthesis with multiple steady states became lower than that in the case in which multiple steady states did not exist because of the improvement in the reaction conversion and the reduction in the reboiler duties of the recovery columns. However, the factor leading to an improvement in the reaction conversion has not been clarified. If the factors leading to an improvement in the reaction conversion can be clarified, this may lead to an improvement in the process performance of a reactive distillation column without the multiple steady-state conditions.

The present work investigated the factors leading to an improvement in the reaction conversion for operation in the multiple steady states. This study focused on the profile of the reactive distillation column used for TAME synthesis. In our previous study, the multiple steady states did not exist for a reflux ratio of 1, but did exist for a reflux ratio of 2. The column profiles of the steady-state solutions for reflux ratios of 1 and 2 were compared. A simulation model of the reactive distillation column with an intermediate condenser at the top of the reactive section was developed to manipulate the internal flow rate of the reactive distillation column. The effect of the intermediate condenser duty on the amount of TAME product and the reboiler duty of the reactive distillation column was clarified. Furthermore, a simulation model of the reactive distillation column with an intermediate condenser and an intermediate reboiler was developed. The effects of the intermediate cooling and heating on the amount of TAME product and the reboiler duty of the reactive distillation were evaluated. Furthermore, the variation in the process performance of the reactive distillation column when the internal vapor liquid flow rate was manipulated was discussed.

2. Simulation Model

2.1. Reaction Kinetics and Physical Properties

The present study focused on TAME synthesis. TAME is produced from 2-methyl-1-butene (2M1B) and 2-methyl-2-butene (2M2B) with methanol (MeOH). The liquid-phase reversible reactions were considered.

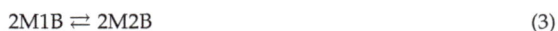

$$2M1B + MeOH \rightleftharpoons TAME \tag{1}$$

$$2M2B + MeOH \rightleftharpoons TAME \tag{2}$$

$$2M1B \rightleftharpoons 2M2B \tag{3}$$

The reaction kinetics for TAME synthesis were proposed by several researchers [10,16]. This study used the simple power law model, with the kinetics parameters for the forward and reverse reactions proposed by Al-Arfaj and Luyben [16]. Appendix A presents the reaction kinetics equation and parameters. The Wilson–RK model was used to predict the physical properties of the mixture [7]. The parameters used to predict the physical properties were applied in the Aspen Plus Database of Aspen Plus (Aspen Technology, Inc., Bedford, MA, USA).

2.2. Reactive Distillation Column

Figure 1 shows a schematic diagram of the reactive distillation column examined herein. The reactive distillation column consists of the rectifying, reaction, and stripping sections. TAME was obtained from the bottom-out stream. Table 1 lists the design and operating conditions of the reactive distillation column [15]. A mixture containing the reactant and an inert component was fed to the 15th stage of the column. In this case, to simplify the analysis, the mixture did not contain TAME. Reflux ratio and reboiler duty were used as the operating variables. The reflux ratio was set to either 1 or 2, while the reboiler duty was varied between 6 and 13 MW. Figure 1b presents a schematic diagram of the reactive distillation column with one intermediate condenser. The intermediate condenser was added to the top of the reactive section (i.e., 28th stage). The intermediate condenser duty ($Q_{\text{inter-condenser}}$) was varied between 0 and −5 MW. Figure 1c depicts a schematic diagram of the reactive distillation column with one intermediate condenser and one intermediate reboiler. The intermediate condenser was added to the top of the reactive section (i.e., 28th stage). The intermediate reboiler was added to the bottom of the reactive section (i.e., 15th stage). The intermediate condenser duty was varied between 0 and −5 MW. The intermediate reboiler duty ($Q_{\text{inter-reboiler}}$) was determined as follows:

$$Q_{\text{inter-reboiler}} = -Q_{\text{inter-condenser}}. \tag{4}$$

The present study did not attempt to optimize the locations of the intermediate condenser and reboiler because our objective was to clarify the factors leading to an improvement in the reaction conversion during operation in the multiple steady states. In addition, the energy input to the intermediate condenser and reboiler was defined as shown in Equation (4) to avoid the influence of any deviation in the energy balance on the process characteristics. A model of the reactive distillation column was developed using the RadFrac model in Aspen Plus.

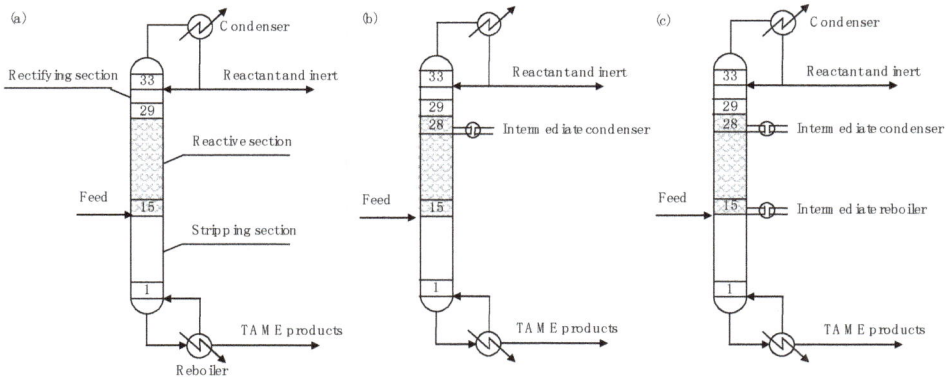

Figure 1. Schematic diagrams of the reactive distillation columns for TAME (*tert*-amyl methyl ether) synthesis: (**a**) conventional, (**b**) added one intermediate condenser, and (**c**) added one intermediate condenser and one intermediate reboiler.

Table 1. Input parameters for the steady-state simulation model of the reactive distillation column.

Design conditions		
Number of stages	33	stages
Number of rectifying stages	5	stages
Number of reactive stages	14	stages
Number of stripping stages	14	stages
Feed location	15th	stage
Amount of catalyst	1100	kg
Bulk density of catalyst	900	kg/m^3
Operating conditions		
Pressure of the top	253	kPa
Pressure drop of stage	1	kPa
Reflux ratio	1, 2	-
Reboiler duty	Variable	MW
Feed conditions		
Flow rate	800	kmol/h
Pressure	280	kPa
Temperature	332	K
Mole fraction		
2-Methyl-butane	0.384	-
1-Pentane	0.029	-
2-Methyl-1-butane	0.022	-
n-Pentane	0.068	-
cis-2-Pentene	0.124	-
2-Methyl-2-butene	0.196	-
Methanol	0.177	-

3. Simulation Results and Discussion

3.1. Process Characteristics

Our previous study reported the condition in which multiple steady states exist, as shown in Figure 3 of previous work [15]. For a reflux ratio of 1, multiple steady states did not exist, but were observed at a reflux ratio of 2 for a reboiler duty between 12.25 and 12.12 MW. This study focused on four steady-state solutions. Table 2 summarizes the simulation results for each steady-state solution. Three solutions with reboiler duty values in the middle of the range were selected for the multiple steady states. The MeOH conversion (ε_{MeOH}) as the reactant was calculated using the MeOH flow rate in the feed (F_{MeOH}), distillate (D_{MeOH}), and bottom (B_{MeOH}):

$$\varepsilon_{MeOH} = \frac{(F_{MeOH} - (D_{MeOH} + B_{MeOH}))}{F_{MeOH}}. \tag{5}$$

The MeOH conversions in the steady-state solutions in the multiple steady states became higher than those in steady state 1. This result indicates that the reaction conversion in the multiple steady states can be improved.

Table 2. Simulation results at various steady-state solutions.

Steady State	Input Variables			Output Variables		
	Reflux Ratio (-)	Reboiler Duty (MW)	$x_{B,TAME}$ (-)	Distillate Flow Rate (kmol/h)	Bottom Flow Rate (kmol/h)	MeOH Conversion (%)
1	1.00	8.25	1.00	690.6	54.68	38.7
2	2.00	12.19	0.67	655.3	86.69	41.1
3	2.00	12.19	0.86	659.0	75.87	46.1
4	2.00	12.19	1.00	656.7	71.64	50.7

This section discusses the factors leading to the reaction conversion improvement in terms of the column profiles. Figures 2–5 show the column profiles for the temperature, vapor, and liquid flow rates, mole fraction, and generated TAME amounts, respectively. From the temperature profile, in the stripping section, the temperature in the 1st stage for steady states 1 and 4 reached 398.4 K, which is the boiling point of TAME. The TAME and MeOH mixture was obtained as shown in Figure 4; hence, the temperatures in the 1st stage for steady states 2 and 3 were 362.7 and 367.4 K, respectively. However, the temperatures in the reactive sections were nearly the same in each of the four steady states; thus, the temperature in the reactive distillation column has little effect on the MeOH conversion.

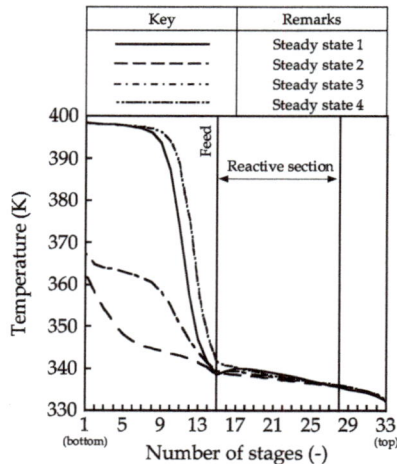

Figure 2. Temperature profiles in the reactive distillation column.

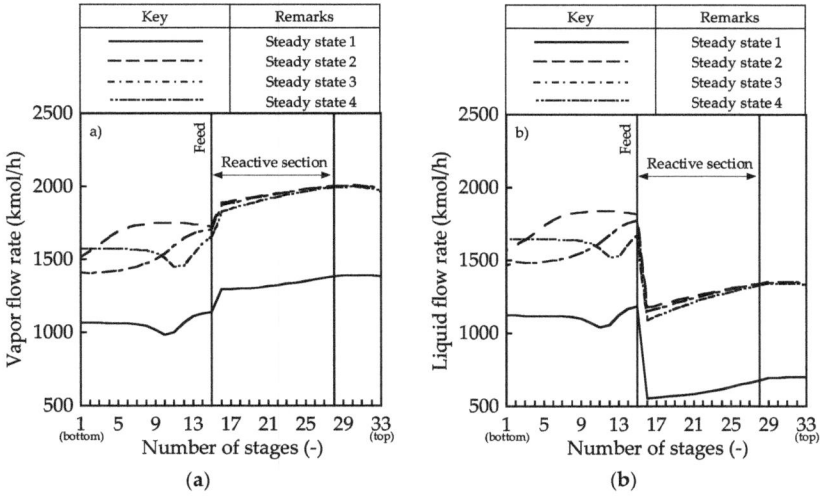

Figure 3. Internal flow rate profiles in the reactive distillation column: (**a**) vapor and (**b**) liquid.

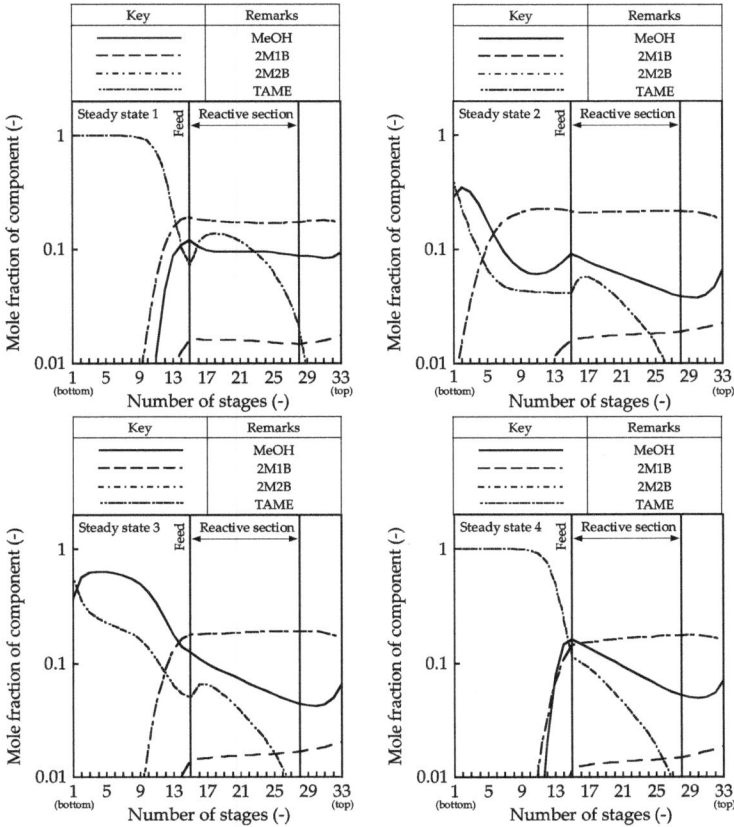

Figure 4. Composition profiles in the reactive distillation column.

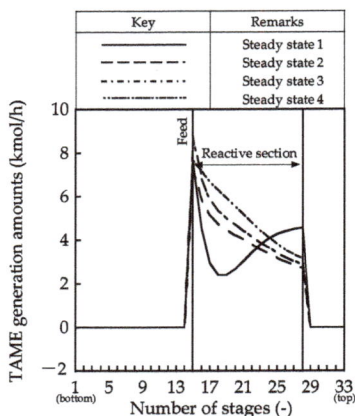

Figure 5. TAME-generated amount profiles in the reactive distillation column.

The vapor and liquid flow rate profiles showed that the vapor and liquid flow rates in steady states 2–4 were larger than that in steady state 1 because the reflux ratio in steady states 2–4 were higher than that in steady state 1. Large vapor and liquid flow rates promoted separation. Therefore, the vapor–liquid flow rate contributed to an increase in the MeOH conversion.

The composition profiles and the generated TAME amount profile showed little change in the mole fractions of MeOH, 2M1B, and 2M2B in the rectifying and reactive sections in steady state 1. The reactants were distilled from the distillate stream; hence, the generated TAME amounts became small. In contrast, in steady states 2–4, the mole fraction of MeOH in the reactive section decreased along with the stage number because MeOH was consumed by TAME synthesis around the bottom of the reactive section. In the stripping section, TAME was not separated from MeOH, and MeOH was discharged from the bottom stream in steady states 3 and 4. By contrast, high-purity TAME was obtained in steady state 5. For these column profiles, the internal vapor–liquid flow rate must be controlled, and the discharge of reactants must be prevented to increase the reaction conversion.

3.2. Effect of an Intermediate Condenser

Based on the abovementioned results, the small MeOH conversion in steady state 1 was caused by the small internal flow rate and the reactant discharge from the top of the reactive section. The internal flow rate may be increased, such that the MeOH discharge is reduced, by condensing the vapors from the top of the reactive section. Thus, a simulation model of the reactive distillation with an intermediate condenser at the 28th stage (top of the reactive section) was developed (Figure 1b). The effect of the intermediate condenser duty on the column profiles was then evaluated. The intermediate condenser duty was changed from 0 to −5 MW. The reboiler duty was adjusted such that the TAME mole fraction in the bottom stream reached 1.00.

Figure 6 shows the effect of the intermediate condenser duty on the bottom flow rate and reboiler duty. The bottom flow rate increased with an increase in the intermediate condenser duty, indicating that the MeOH conversion increased because the TAME mole fraction in the bottom stream was 1.00. Figure 7 presents the vapor–liquid flow rate and the composition profiles for an intermediate condenser duty of −3.0 MW. In this case, the internal flow rate was larger than that in steady state 1. The composition profile was such that the mole fraction of MeOH decreased along with the stage number in the reactive section, leading to an improvement in the reaction performance. However, in the stripping section, the mole fraction of MeOH was higher than that in steady state 3. The MeOH conversion may increase if the amount of MeOH discharged from the bottom of the reactive section decreases.

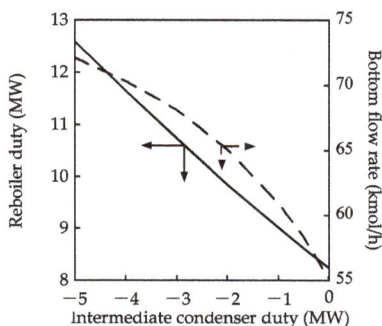

Figure 6. Effect of the intermediate condenser duty on the reboiler duty and bottom flow rate.

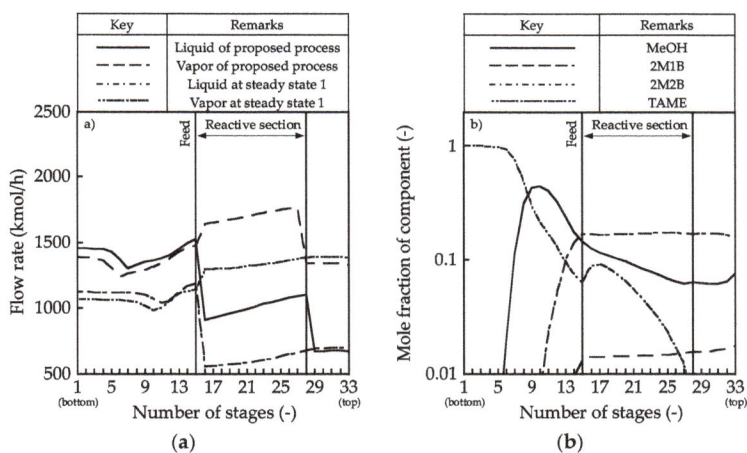

Figure 7. Column profiles in the reactive distillation column with one intermediate condenser: (**a**) vapor and liquid flow rate and (**b**) composition.

3.3. Effects of the Intermediate Condenser and the Intermediate Reboiler

This study also examined liquid vaporization from the bottom of the reactive section with the intermediate reboiler to reduce the amount of MeOH discharged from the bottom of the reactive section. A simulation model for the reactive distillation column with an intermediate reboiler and condenser was developed (Figure 1c).

Figure 8 shows the effects of the intermediate condenser and reboiler duty on the bottom flow rate and reboiler duty. The reboiler duty was adjusted such that the TAME mole fraction in the bottom stream reached 1.00. The effect of the intermediate reboiler duty on the bottom flow rate was similar to that in Figure 6. The degree of MeOH conversion also cannot increase beyond the results attained with reactive distillation when using an intermediate condenser. However, the reboiler duty can be reduced with the addition of the intermediate reboiler presumably because the separation of TAME and MeOH occurs as a result of introducing the intermediate reboiler.

Finally, the energy consumption for each reactive distillation column is discussed. A comparison was made with the energy consumption when TAME was produced at a rate of 72 kmol/h from the bottom-out stream. The energy consumptions in steady state 4 of the reactive distillation column with one intermediate condenser and of the reactive distillation column with one intermediate condenser and reboiler were 612, 628, and 637 kJ/mol. The present study did not attempt to optimize either the locations of the intermediate condenser and reboiler or the energy input. Further energy savings could possibly be achieved by optimizing these locations and the input energy.

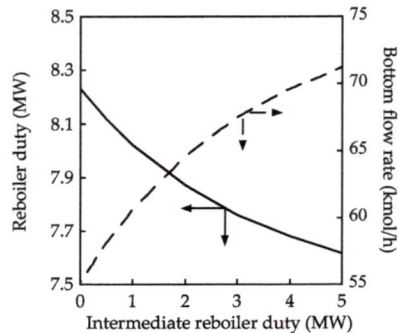

Figure 8. Effect of the intermediate reboiler duty on the reboiler duty and bottom flow rate.

4. Conclusions

The present study clarified the factors leading to an improvement in the reaction conversion during operation in multiple steady states for the reactive distillation column used for TAME synthesis. The column profiles for reflux ratios of 1 (no multiple steady state) and 2 (multiple steady states) were compared. The comparison of the column profiles for four steady-state solutions clearly showed that the internal vapor and liquid flow rates in the reactive distillation column affected the reaction conversion. An intermediate condenser was introduced at the top of the reactive section to control the liquid flow rate in the reactive distillation column. The effect of the intermediate condenser duty on the TAME production was evaluated, with the amount of TAME product increasing from 55.2 to 72.1 kmol/h as a result of operating the intermediate condenser between 0 and −5 MW. The liquid vaporization from the bottom of the reactive section with the intermediate reboiler was also examined. As a result, the amount of TAME product cannot increase beyond the amounts produced by reactive distillation using an intermediate condenser, but the reboiler duty can be reduced through the addition of the intermediate reboiler. These results indicate that the liquid and vapor flow rates in the reactive distillation column influence the reaction and separation performances, respectively.

In the future, we will investigate the attainment of optimal vapor–liquid flow rates with small input energy. The temperature difference between the top and the bottom of the reactive section was 2.9 K (Figure 2). Heat integration between the intermediate condenser and the intermediate reboiler through the application of the heat-pump technology may be effective [17–19].

Author Contributions: Conceptualization and methodology: T.Y. and K.M.; validation: T.Y., D.N.-R., and H.M.; investigation: T.Y. and K.M.; resources: K.M.; writing and original draft preparation: T.Y.; writing, review, and editing: K.M., D.N.-R., and H.M.; and supervision: K.M.

Funding: This research received no external funding.

Conflicts of Interest: The authors declare no conflict of interest.

Appendix A

The following are the reaction kinetics equations and the reaction parameters [16]:

$$R_1 = A_f e^{-E_f/RT} x_{2M1B} x_{MeOH} - A_b e^{-E_b/RT} x_{TAME} \tag{A1}$$

$$R_2 = A_f e^{-E_f/RT} x_{2M2B} x_{MeOH} - A_b e^{-E_b/RT} x_{TAME} \tag{A2}$$

$$R_3 = A_f e^{-E_f/RT} x_{2M1B} - A_b e^{-E_b/RT} x_{2M2B}. \tag{A3}$$

Table A1. Reaction kinetics parameters.

Reaction	A_f kmol/(s·kg)	E_f kJ/mol	A_b kmol/(s·kg)	E_b kJ/mol
R_1	1.3263×10^8	76.1037	2.3535×10^{11}	110.5409
R_2	1.3718×10^{11}	98.2302	1.5414×10^{14}	124.994
R_3	2.7187×10^{10}	96.5226	4.2933×10^{10}	104.196

References

1. Stankiewicz, A.I.; Moulijn, J.A. Process intensification: Transforming chemical engineering. *Chem. Eng. Prog.* **2000**, *96*, 22–34.
2. Cárdenas-Guerra, J.C.; López-Arenas, T.; Lobo-Oehmichen, R.; Pérez-Cisneros, E.S.A. Reactive distillation process for deep hydrodesulfurization of diesel: Multiplicity and operation aspects. *Comput. Chem. Eng.* **2010**, *34*, 196–209. [CrossRef]
3. Subawalla, H.; Fair, J.R. Design guidelines for solid-catalyzed reactive distillation systems. *Ind. Eng. Chem. Res.* **1999**, *38*, 3696–3709. [CrossRef]
4. Taylor, R.; Krishna, R. Modeling reactive distillation. *Chem. Eng. Sci.* **2000**, *55*, 5183–5229. [CrossRef]
5. Huss, R.S.; Chen, M.; Malone, M.F.; Doherty, M.F. Reactive distillation for methyl acetate production. *Comput. Chem. Eng.* **2003**, *27*, 1855–1866. [CrossRef]
6. Huang, K.; Wang, S.J. Design and control of a methyl tertiary butyl ether (MTBE) decomposition reactive distillation column. *Ind. Chem. Eng. Res.* **2007**, *46*, 2508–2519. [CrossRef]
7. Mohl, K.D.; Kienle, A.; Gilles, E.D.; Rapmund, P.; Sundmacher, K.; Hoffmann, U. Steady-state multiplicities in reactive distillation columns for the production of fuel ethers MTBE and TAME: Theoretical analysis and experimental verification. *Chem. Eng. Sci.* **1999**, *54*, 1029–1043. [CrossRef]
8. Jacobs, R.; Krishna, R. Multiple solutions in reactive distillation for methyl *tert*-butyl ether synthesis. *Ind. Eng. Chem. Res.* **1993**, *32*, 1706–1709. [CrossRef]
9. Nijhuis, S.A.; Kerkhof, F.P.J.M.; Mak, A.N.S. Multiple steady states during reactive distillation of methyl *tert*-butyl ether. *Ind. Eng. Chem. Res.* **1993**, *32*, 2762–2774. [CrossRef]
10. Baur, R.; Taylor, R.; Krishna, R. Bifurcation analysis for TAME synthesis in a reactive distillation column: Comparison of pseudo-homogeneous and heterogeneous reaction kinetics models. *Chem. Eng. Process.* **2003**, *42*, 211–221. [CrossRef]
11. Wang, J.; Change, Y.; Wang, E.Q.; Li, C.Y. Bifurcation analysis for MTBE synthesis in a suspension catalytic distillation column. *Comput. Chem. Eng.* **2008**, *32*, 1316–1324. [CrossRef]
12. Ramzan, N.; Faheem, M.; Gani, R.; Witt, W. Multiple steady states detection in a packed-bed reactive distillation column using bifurcation analysis. *Comput. Chem. Eng.* **2010**, *34*, 460–466. [CrossRef]
13. Jairne-Leal, J.E.; Bonilla-Petriciolet, A.; Segovia-Hernandez, J.G.; Hernandez, S.; Hernandez-Escoto, H. On the multiple solution of reactive distillation column for production of fuel ethers. *Chem. Eng. Process.* **2013**, *72*, 31–41. [CrossRef]
14. Yamaki, T.; Shishido, M.; Matsuda, K.; Matsumoto, H. Effect of operating conditions on the multiple steady states in reactive distillation. *Kagaku Kogaku Ronbunshu* **2014**, *40*, 244–249. [CrossRef]
15. Yamaki, T.; Matsuda, K.; Duangkamol, N.R.; Matsumoto, H. Energy-saving performance of reactive distillation process for TAME synthesis through multiple steady state condition. *Chem. Eng. Process.* **2018**, *130*, 101–109. [CrossRef]
16. Al-Arfaj, M.A.; Luyben, W.L. Plantwide control for TAME production using reactive distillation. *AIChE J.* **2004**, *50*, 1462–1473. [CrossRef]
17. Kansha, Y.; Tsuru, N.; Fushimi, C.; Tsutsumi, A. Integrated process module for distillation processes based on self-heat recuperation technology. *J. Chem. Eng. Jpn.* **2010**, *43*, 502–507. [CrossRef]

18. Kiss, A.A.; Landaeta, S.J.F.; Ferreira, C.A.I. Towards energy efficient distillation technologies—Making the right choice. *Energy* **2012**, *47*, 531–542. [CrossRef]

19. Matsuda, K.; Iwakabe, K.; Nakaiwa, M. Recent advances in internally heat-integrated distillation columns (HIDiC) for sustainable development. *J. Chem. Eng. Jpn.* **2012**, *45*, 363–372. [CrossRef]

![processes logo] **processes**

MDPI

Article

Optimization of Reaction Selectivity Using CFD-Based Compartmental Modeling and Surrogate-Based Optimization

Shu Yang, San Kiang, Parham Farzan and Marianthi Ierapetritou *

Department of Chemical and Biochemical Engineering, Rutgers, The State University of New Jersey, 98 Brett Road, Piscataway, NJ 08854, USA; yang.shu.public@gmail.com (S.Y.); san.kiang@gmail.com (S.K.); parhamfarzan@gmail.com (P.F.)
* Correspondence: marianth@soe.rutgers.edu; Tel.: +1-848-445-2971

Received: 19 November 2018; Accepted: 24 December 2018; Published: 29 December 2018

Abstract: Mixing is considered as a critical process parameter (CPP) during process development due to its significant influence on reaction selectivity and process safety. Nevertheless, mixing issues are difficult to identify and solve owing to their complexity and dependence on knowledge of kinetics and hydrodynamics. In this paper, we proposed an optimization methodology using Computational Fluid Dynamics (CFD) based compartmental modelling to improve mixing and reaction selectivity. More importantly, we have demonstrated that through the implementation of surrogate-based optimization, the proposed methodology can be used as a computationally non-intensive way for rapid process development of reaction unit operations. For illustration purpose, reaction selectivity of a process with Bourne competitive reaction network is discussed. Results demonstrate that we can improve reaction selectivity by dynamically controlling rates and locations of feeding in the reactor. The proposed methodology incorporates mechanistic understanding of the reaction kinetics together with an efficient optimization algorithm to determine the optimal process operation and thus can serve as a tool for quality-by-design (QbD) during product development stage.

Keywords: mixing; CFD-simulation; surrogate-based optimization; compartmental modeling; competing reaction system; optimization; model order reduction

1. Introduction

In chemical synthesis, many important reactions can be accompanied by undesired side-reactions. This leads to wastes and affects product quality. Therefore, incorporating knowledge of mixing can substantially improve reaction selectivity and yield, in addition to enhancing process safety. Furthermore, due to a growing variety of reactors, characterization of mixing has become vital in process development [1,2]. To achieve optimal selectivity and yield, appropriate modeling of the mass transport process in reactors is critical [3]. Nevertheless, owing to the complexity of mass transport in turbulent flow, analyzing mixing remains a difficult problem.

Frequently, mixing in reactor is approximated by residence time distribution (RTD) analysis, where residence time is experimentally measured through a tracer test. This approximation has been proved to work relatively well, however *"RTD is not a complete description of structure for a particular reactor or system of reactors"* [3]. Therefore, when reaction with high conversion rates are considered, analysis solely based on RTD can lead to significant error [4]. Based on local sensors, RTD characterization of tracer test can be improved by mixing time measurement [5,6]. However, considering potential bias and the requirement for specific equipment for tracer tests, RTD and mixing time measurement have become less preferable comparing to the more resource-effective benchmark reaction method [7]. Therefore, competitive reaction systems have become the standard for mixing analysis [7–9]. Among the competitive reaction systems, the *"well-documented and highly reliable"* [7] Bourne reaction and Villermaux reaction is most commonly adopted [10]. Nevertheless, benchmark reaction tests provide only the input-output relationship, without detailed description of the process dynamics. Therefore, developing optimal process operation experimentally remains challenging.

With rapid advances in computer technology, computational fluid dynamics (CFD) has become a powerful tool to study mixing. Comparing with the experimental methods, it provides detailed understanding of mixing phenomenon in a timely-efficient manner without requiring meticulous choice of equipment and sensors. In reaction engineering, CFD has been employed to study mixing in chemical reactors and bio-reactors. Mixing of liquid-liquid system [11], solid-liquid system [12] or non-Newtonian fluid [13] are studied with CFD, which agrees well with experimental data. In the presence of complex chemical reactions [14] or biological metabolism [15] CFD is implemented to provide detailed understanding of the mass transfer process. Complex hydrodynamics, mass transfer, heat transfer and reaction kinetic can be satisfactorily captured by CFD, making it a powerful tool in process design.

As a result, CFD has been implemented to optimize mixing by improve reactor design. Studies reported in the literature mainly focus on improving geometrical attributes of reactors. Researches that directly integrate detailed CFD simulation with optimization algorithms has successfully improved reactor design [16–20]. Due to the complexity of CFD simulations, the chosen optimization algorithms are often meta-heuristic, such as genetic algorithm (GA) [16–18] and particle swarm methodology [20]. They can address complex black-box problems, such as reactor geometry [16,17] or impeller configuration [18], but require a large number of function evaluations, which leads to a large number of computationally expensive CFD runs. To improve the computational efficiency of direct CFD simulations, hybrid methods have been proposed that replace direct CFD evaluation with simpler data-driven models [21–25], which are then integrated with GA. Successful implementation have been reported in the literature that use neural network [22,23], Gaussian process [24] and radial basis function (RBF) [25]. However, building confidence in those data-driven models requires large number of CFD runs, and balancing computational efficiency with accuracy is non-trivial [26].

Apart from improving mixing through optimized design, to the best of the authors' knowledge, there is no work that improves mixing by optimizing dynamic process operation based on CFD simulations. The main reasons for the limited implementation is the complexity of CFD simulations. To optimize dynamic process operation, more decision variables should be considered. Since GA would suffer from "curse of dimensionality" [27], significantly more CFD runs would be required, leading to increasing computational expense and degrading performance.

In this work, a framework is developed trying to leverage CFD simulations to optimize process operation. Firstly, CFD-based compartmental model is built to replace direct CFD simulations. Comparing to the data-driven models implemented for reactor geometry optimization, compartmental models are finite-volume physical models, which provide satisfactory accuracy while requiring significantly less CFD runs. Secondly, a surrogate-based optimization algorithm using radial-basis function is implemented, which has been proven to be more efficient than GA [28]. This work offers a compact and systematic framework for improving reaction selectivity with a numerically efficient Quality-by-Design (QbD) tool. In a case study, Bourne reaction is employed as a benchmark, which serve as a more explicit quantification for the efficiency of mixing. The proposed framework is compared with process operations optimized based on ideal mixing model, suggesting great improvement by leveraging CFD simulations. By integrating CFD-based compartmental model with surrogate-based optimization, the proposed framework has shown a great potential for fast process development.

2. Integrating CFD-Based Compartmental Model with Surrogate Based Optimization

This section presents the development and implementation of the proposed methodology. Initially, a detail description of flow field in the reactor is generated by a CFD simulation. It should be noted that in stirred tank reactors the flow field are considered independent from chemical reactions. As a result, the fluid dynamics data can be used for different reactions, which could contribute to rapid process design and cost reduction. The result from CFD simulation is used to develop a compartmental model, which would be discussed in Section 2.1. Comparing to direct CFD simulation, computational complexity of compartmental model is significantly reduced. Therefore, the optimal process design can be determined numerically without prohibitive computational expense. The process selectivity is then optimized by integrating the compartmental model with surrogate-based optimization, as will be discussed in Section 2.2.

2.1. CFD-Based Compartmental Model

2.1.1. A Brief Review of Compartmental Model

Compartmental model defines a matrix of perfectly mixed control volumes interconnected by the exchanging mass flux. In this method the reactors are discretized into a set of homogeneous control volumes according to the defined mesh. The volume-averaged variable in all control volumes are solved together to represent the space distribution inside the reactor. Compartmental modeling was regarded as a crude tool to study transport process and only provide basic understandings [29]. However, by incorporating detailed CFD simulation to compartmental model, substantial improvement can be achieved leading to satisfactory agreement with experimental data without loss of computational efficiency [30,31].

Considering the excessive computational and economical expense usually required by CFD simulation, for Chemistry, Manufacturing, and Controls (CMC) development, CFD-based compartmental model have been adopted as computationally cheaper alternative [4,30,32–34] In addition to the reduced computational expense, CFD-based compartmental model provides the required simplifications for development work. Unlike CFD, which is not widely available and requires special know-how, CFD-based compartmental model can be easily implemented for different reaction systems based on flow field data determined beforehand. Adjustment in chemical kinetics do not usually require extra CFD simulation, which could save time and reduce cost.

In this proposed methodology, compartmental model is developed from CFD simulation based on the idea outlined by Bezzo et al. [33]. Two key steps are required for model construction: (1) Topological mapping between two models through aggregating CFD cells into compartments. (2) Quantifying mass flux between different compartments. Topology mapping between CFD and compartmental model can be done either manually or automatically. Manual allocation of CFD cells is based on preliminary knowledge of flow field, which can be conducted prior to CFD simulation [15,31,35,36]. Automatic mapping, on the other hand, merges computational cells based on CFD simulation to form meaningful homogeneous control volumes [4,37]. Manual allocation usually leads to simpler mesh structure, which allow for efficient implementation of optimization tools. Therefore, in this work manual allocation is conducted, as will be outlined in sub Section 2.1.3.

2.1.2. Compartmental Model Development

Mixing of particles inside the reactor is described by Equation (1), where c_i is the concentration of species i, N_i represents the mass flux of species i, and R_i denotes the source of species i. Compartmental models are obtained by volume averaging Equation (1) over each predefined compartment V, as described in Equation (2).

$$\frac{\partial C_i}{\partial t} = -\nabla \cdot N_i + R_i,$$ (1)

$$\frac{d}{dx} \int_V \overline{C_i} dV = - \int_V \nabla \cdot N_i dV + \int_V R_i dV.$$ (2)

Adopting the divergence theorem and replacing C_i with the volume-averaged concentration $\overline{C_{i,K_j}}$, Equation (2) is modified to Equation (3), where V_j is the volume of control volume K_j, and S_j is the surface of control volume K_j. The mass flux N consist of convection and diffusion, where the diffusion is modelled by Fick's law with diffusion coefficient D. The source term R_i models the homogeneous consumption and generation of species i, which include chemical reactions and micro mixing. Due to the assumption of homogeneous compartments, the volume integral of source term in Equation (5) is modified as follows:

$$V_j \frac{d\overline{C_{i,K_j}}}{dx} + \oint_{S_j} N_{i,K_j} \cdot n dS = \int_{K_j} R_i dV,$$ (3)

$$V_j \frac{d\overline{C_{i,K_j}}}{dx} = -- \int_{S_j} v \cdot \overline{C_{i,K_j}} \cdot n \, dS + \int_{S_j} D \cdot \nabla C \cdot n \, dS + \int_{K_j} R_i dV,$$ (4)

$$\int_{K_j} R_{i,K_j} = V_j \cdot R_{i,K_j}.$$ (5)

It should be recognized that the homogeneous assumption depends upon the Damköhler number (Da) in each compartment, which is a strong function of grid size, as will be discussed in sub Section 2.1.3. Moreover, in this work diffusion mass transfer between different compartments are neglected, which is also justified in sub Section 2.1.3 based on an analysis of the Péclet number (Pe). The dominance of convective mass transfer would simplify 4 into Equation (6), where Q_{jk} denotes the flow rate from control volume j to control volume k.

$$V_j \frac{d\overline{C_{i,K_j}}}{dt} = -\overline{C_{i,K_j}} \sum_k Q_{j,k} + \sum_l \left(Q_{l,j} \cdot \overline{C_{i,K_j}} \right) + V_j \cdot R_{i,K_j}.$$ (6)

Equation (6) is a finite-volume mixing model parameterized by mass flow rate Q and compartment volumes V. Since CFD is also based on fine-volume models, parameters of this mixing model can be effectively determined from CFD simulations, based on the topology mapping and merging strategy proposed by Bezzo et al. [33].

Firstly, a steady-state CFD simulation should be developed and calibrated to yield a good prediction of flow field inside the reactor. Then the computational cells of CFD mapped to the same compartments are group into ensembles. Cells in the same ensemble are aggregated together, where, the volumes and mass flow rate and summed together to determine the parameters of each compartment. To achieve a good balance between computational complexity and model accuracy, the resolution of the compartmental model is determined based on a grid independence test as will be discussed in sub Section 2.1.3. For illustration purposes, a well-known pair of parallel competitive Bourne reactions is studied. As will be discussed later in the case study, this reaction system is composed of a first-order decay and a parallel second order coupling reaction [38].

2.1.3. Grid Independence

From numerical perspective, compartmental model is an upwind discretization of mass balance equation with finite volume method. Underlying this discretization scheme lies the assumption of homogeneity inside each control volume. Therefore, compartmental model would exhibit higher diffusivity than the true medium. The deviation caused by compartmentalization depends on the system being modelled and the type of discretization that is used.

One heuristic rule is that with higher resolution grid, the discretized model should behave more like the continuous case. However, with increasing resolution, the complexity of the model also increases, which leads to higher computational expenses. Moreover, decreasing grid size would lead to a decreasing Péclet number, which would undermine the assumption of ignoring diffusion mass transfer. Therefore, a grid independence analysis should be conducted to find the optimal grid density to map the CFD data to compartmental model.

In this work the grid independence test is performed in two steps. First the Damköhler number (Da) and Péclet number (Pe) are analyzed, as suggested in Equations (7) and (8), which are critical to justify the compartmentalization of model discussed in sub Section 2.1.2.

$$Da = \frac{k_1 C_A + k_2 C_b C_A}{u C_A / L} < 1, \tag{7}$$

$$Pe = \frac{Lu}{D} > 1. \tag{8}$$

This analysis determines the lower and upper bounds of length scale for the compartments, which could serve as a starting point for the grid independence test. Then the initial choice of length scale is improved in an iterative manner. Simulations based on compartmental models with decreasing length scale are tested. When the simulation results are no longer changing with the increasing grid density, the model resolution can be considered as sufficient.

It should be mentioned that the optimal grid density depends on reaction kinetics. If the time constant of chemical reactions is significantly larger than that of mixing, this process could be considered mixing-insensitive and perfect mixing assumption could be adopted without harming model precision. By contrast, for fast reaction, the deviation caused by perfect-mixing assumption could be significant, which require higher resolution. If the characteristic time scale of reaction is in orders of magnitude less than micro-mixing, which is in the order of 10^{-3} s [11], micro-mixing would dominate chemical reaction. As a result, the rate law of chemical reaction should be replaced with micro-mixing models. Although for different reaction kinetics we can use the same steady state solution from CFD, if new reaction kinetics are used, grid independence test should be conducted with the updated model.

2.2. Surrogate-Based Optimization

2.2.1. A Brief Review of Surrogate-Based Optimization

Surrogate-based optimization have been the focus of interest in the derivative-free optimization literature. Commonly seen in science and engineering studies are complex computer simulations and experiments conducted to gain understanding of systems. As a result, for these problems derivative information is either unavailable or prohibitively hard to get, making it impossible to implement deterministic optimization methods efficiently [39]. Therefore, there is a high interest in developing methods to handle the optimization problems where limited or noisy information is available [40].

Surrogate-based optimization use surrogate models, which are simpler models that can mimic complex phenomenon, to guide the search in derivative-free optimization problems. Since surrogate models are computationally less demanding, surrogate-based optimization is a good compromise between describing the complex process and remaining computationally feasible. It has been demonstrated that surrogate-based optimization displays superior performance for derivative-free optimization problems [41]. Most popular surrogate models implemented for optimization methods are RBF [42–46] and Kriging [47–51], due to their capability to provide prediction uncertainty. Artificial neural networks (ANN) have excellent fitting characteristics with low complexity, therefore implementations of ANN for surrogate-based optimization (SBO) is popular for various engineering applications [22,52–54].

SBO works in an iterative manner. In the initial step, several sampling point are chosen, and an initial surrogate model is built based on function evaluation at those sampled points. Then new sampling points are determined by evaluating the surrogate model. At new sampling point, the original model is evaluated and the surrogate model is updated. This process is conducted iteratively until a stopping criterion is met, and the best design point is chosen. In this work, the mixed-integer optimization problem is solved with SBO based on the work of Müller [55].

2.2.2. Problem Formulation

In this work, the location and rate of feeding are optimized to improve reaction selectivity. Due to the perfect mixing assumption of control volumes, feeding location is represented by the index of compartment it resides at. It is worth noting that while the feeding location should be fixed throughout the process, the feeding rate could change dynamically. Therefore, by taking advantage of the extra degree of freedom through adopting a changing feeding rate, reaction selectivity could be further improved comparing to a fixed rate feeding, as will be discussed in Section 3.5.2. Dynamic profile of feeding rate is defined by splitting the whole process time into N_s discrete feeding stages. Feeding rate is kept constant in each stage, but different feeding rate are employed for different stages. Each feeding stage m is specified by its duration t_m and the adopted feeding rate f_m, which are not defined a priori, instead they are determined by solving an optimization problem.

In order to solve for the optimal operating policy, reaction selectivity should be quantified based on analyzing product distribution. The most intuitive definition is by segregation index, which is based on the ratio of raw material consumed by the desired product to the total raw materials injected. This method was widely adopted in previous work [2,56,57], where the influence of feeding rate was not investigated. However, adopting segregation index as an objective function in this work could lead to trivial solutions, due to the fact feeding rate usually contributes monotonically to product ratio. Without considering the economics of the process, solely focusing on the product ratio would lead to unsatisfactory process design. Thus, it is recommended to use revenue as a way to capture and optimize reaction selectivity. To maximize revenue of chemical processes, the optimization problem is defined as followed.

$$Maximize_{\,n,t_i,f_i} \left(\sum P_R y_R - \sum P_A y_A \right), \tag{9}$$

Subject to:

$$[y_R, y_A] = \varphi \left(n, t_1, f_1, t_2, f_2, \ldots, t_{N_p}, f_{N_p} \right), \tag{10}$$

$$\sum t_m = T, \ \forall m = 1, \ldots, N_s, \tag{11}$$

$$t_m, f_m \geq 0, \ \forall m = 1, \ldots, N_s, \tag{12}$$

$$n \in [0, N_c], \tag{13}$$

where P_R denotes the price of desired product R while y_R denotes its yield. P_A represents the unit cost of raw material A and its consumption is denoted as y_A. Both y_R and y_A are calculated through the simulation φ based on the compartmental model. Addition point n and addition rate profile which is defined by $t_1, f_1, t_2, f_2 \ldots t_{Ns}, f_{Ns}$ are parameters of this simulation. The first set of constraints describe the simulation based on the compartmental model. The second constraint represents that the total time span of all stages is pre-defined as T.

Notice that the number of stages is introduced as a parameter instead of decision variable. This is based on the difficulty of penalizing the monotonic increase of the number of stages. By allowing extra degrees of freedom, an increasing number of stages is always preferred. Reaction selectivity would always benefit from higher degree of freedom provided by the increasing N_p, unless computational expense of solving this optimization problem is taken into consideration. However, this is beyond the scope of this paper. The duration of process T is defined as a parameter, which is usually determined in the production scheduling stage. It is recommended to define T similar to the timescale of mixing in mixing controlled processes to maximize time efficiency of reactors.

3. Case Study

In this section, a case study of a semi-batch process inside a dual-impeller stirred tank reactor is studied. The duration of the whole process is 150 s, in which a fed-batch process is analyzed and optimized. A well-known pair of competitive reactions [38] is introduced to study the influence of mixing on reaction selectivity. The overall objective for the optimization problem is to maximize process productivity, which is defined as the revenue from selling the products minus the total cost of raw materials injected. In this case study, process designed according to CFD-based compartmental model and perfect-mixing model are compared to illustrate the effectiveness of this methodology. Furthermore, constant feeding rate design is compared with time-varying feeding rate to demonstrate that this framework can further improve reaction selectivity by enabling dynamic design.

3.1. Reactor Setup

This study was carried out in a 74 L baffled stirred vessel agitated with a Rushton impeller and a pitched blade turbine, as illustrated in Figure 1. The diameter of the vessel is 0.5 m, and the liquid level is 0.4 m from the bottom of the vessel. The Rushton impeller is assembled 0.14 m below the pitched blade turbine, whose blade angle is 45°. The agitation system is operated at 12 rpm anticlockwise, which drives fluid downwards from the pitched blade turbine to the Rushton impeller.

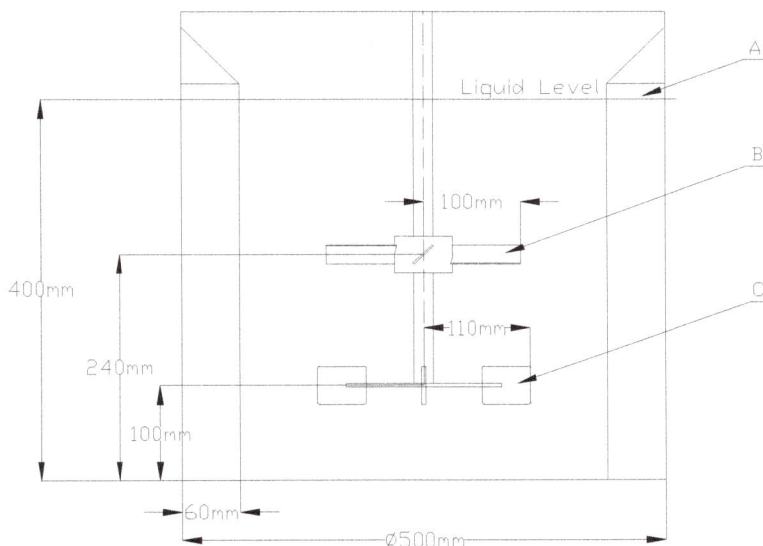

Figure 1. Geometrical dimension of the two-impeller stirred tank. Where (**A**) stands for the baffles, (**B**) represents the pitched blade impeller and (**C**) denotes the Rushton impeller.

3.2. Chemical Kinetics

To study the influence of mixing on reaction selectivity, a well-documented pair of parallel competitive Bourne reactions is integrated. This reaction system is composed of a first-order decay and a parallel second order coupling, where A is a diazonium salt (diazotized 2-chloro-4-nitroaniline) and B is pyrazolone (4-sulphophenyl-3-carboxypyrazol-5-one). R denotes the desirable product which is a dyestuff, and S is the unwanted product of decomposition. The rate constants at 40 °C are $k_1 = 10^{-3}$ s^{-1} and $k_2 = 7000$ m^3 kmol^{-1} s^{-1} at a PH = 6.6 [38]. Both reactants are dissolved in aqueous solution.

$$A \overset{k_1}{\to} S, \tag{14}$$

$$A + B \overset{k_2}{\to} R \tag{15}$$

The vessel is initially charged with pyrazolone solution with a concentration of 1×10^{-3} M. diazonium solution is added into the stirred tank in a semi-batch manner, the concentration of which is 7.4×10^{-1} M.

In this reaction system, the advantage of defining objective function in the form of productivity is pronounced. Considering that the desired reaction happens faster than side reactions, infinitely slow feeding would always be preferred if we want to maximize the ratio between desired product and side product. Based the time scale of mixing, the duration of process is set as 150 s.

3.3. Flow Field Simulation

CFD simulation is adopted to solve for velocity field inside this reactor based on the physical property of the solvent. The influence of the feeding pipe over the flow pattern is ignored. Since the flow rate of injection pipe is 10^2 order smaller than that of the bulk flow inside the reactor, influence of reagent injection over the flow pattern is ignored.

A steady state CFD simulation is conducted with the commercial code of Ansys Fluent 16.0 [58]. The Reynolds-Averaged-Navier-Stocks (RANS) equation was numerically solved with multi-reference frame (MRF) method. To close the equations, k-epsilon turbulence model with standard wall functions was adopted. The velocity field is shown in Figure 2.

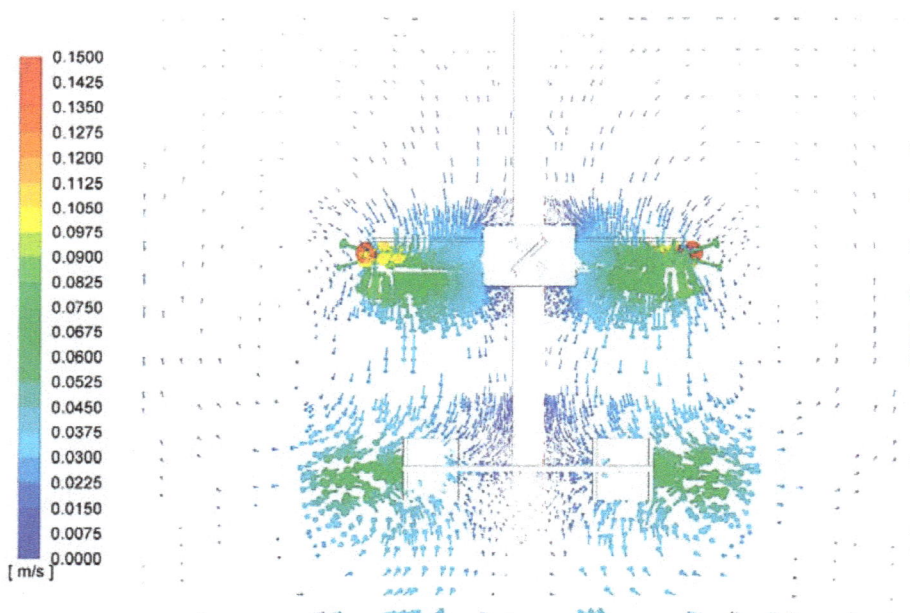

Figure 2. Simulated vector plot of velocity field inside the stirred tank (m/s).

It should be noted that it is necessary to calibrate CFD simulations so that can be confidently utilized. Since this case study is used for demonstration purposes, due to the lack of experimental data, validation of the numerical simulation is not conducted. This will be addressed in future work where this method is applied.

3.4. Compartmental Modeling and Grid Independence Test

To develop compartmental models from steady state CFD simulation, computational cells extracted from CFD are aggregated based on a predefined grid. The grid is defined by evenly dividing the reactor in radial, axial and angular directions. The grid density of each direction is determined based on the grid sensitivity test proposed in Section 2.1.3.

To justify the perfect-mixing assumptions in each compartment, local Damköhler number (Da) is analyzed. As suggested Figure 2, the bulk velocity inside the stirred tank is in the order of 10^{-2} m/s. Based on Equation (7), the upper bound of length scale in each compartment should be 1 m. Furthermore, to justify neglecting diffusive mass transfer, Péclet number (Pe) is analyzed according to Equation (8). Since diffusion coefficient in aqueous solutions are in the order of 10^{-9} m^2/s, the lower bound of compartment length scale is 10^{-7} m. It can be concluded that since in single phase turbulent flow convective mass transfer rate is usually several orders of magnitude higher than that of diffusion, compartmental model can be safely adopted in most single phase stirred tank reactors.

Starting from the upper bound indicated by the analysis of Damköhler number, length scale of the compartments is decreased to test the optimal grid density as discussed in Section 2.1.3. In this work, grid independence test is performed by simulating the injection of diazonium at the top free surface of liquid near the wall. Considering that the injection should be fast enough to show mixing effect, but not excessively fast so that the pyrazolone is instantly depleted and the mixing-sensitive coupling is dominated by the decaying, the feeding rate of diazonium solution is set as 0.5 mL/s, scaled from the work of Nienow [38]. The distribution of different chemical species at the end of process is predicted and monitored. The total number of compartments used for the grid sensitivity test varied from 384 (8 × 8 × 6: axial × radial × angular) to 2352 (12 × 14 × 14). The predicted amount of chemical species varies with the number of compartments and approaches asymptotic values as shown in Figure 3. In good agreement with the scaling analysis, convergence is achieved with 1920 (12 × 16 × 10) compartments for all species, which corresponds to Da < 0.1. For fast model development, Damköhler number can served as an efficient criterion for defining grids [30]. Further results presented in this paper are based on this discretization scheme.

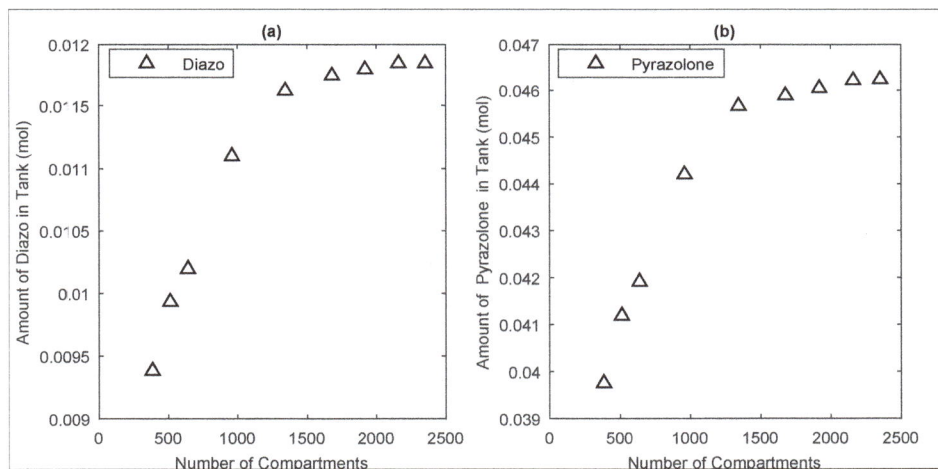

Figure 3. Convergence of predicted (**a**) Diazonium and (**b**) Pyrazolone distribution with number of compartments.

3.5. Optimization and Results

The overall objective for the optimization problem is to maximize process productivity, which is defined based on the price of different materials. In this case study the price of desired product is assumed to be ten times as much as the price of diazonium, which is 10^4 \$/mol. The prices have profound influence on the optimal operating policy. Feeding points are defined with 3 integer variables, representing the corresponding radial, axial and angular index.

The optimization algorithm is first solved for the optimal static operating condition in which reagent is injected in a constant rate. To further improve the process productivity, dynamic operating conditions where the feeding rate changes dynamically are studied and optimized. Dynamic policies comprised of 2 and 3 feeding stages are discussed. Furthermore, by optimizing process design with perfect-mixing assumption, traditional design is compared with this proposed methodology.

3.5.1. Optimal Location of Feeding

In this section, the influence of feeding location on mixing and reaction selectivity is studied. Two operating conditions with different feeding locations are compared; one is at the bottom corner of the reactor while the other one is at the tip of Rushton impeller. The feeding rate (1 mL/s) is kept constant throughout the simulation. In Figure 4 the yield trajectories of desired product are displayed when different feeding locations are adopted. Considering that stronger convective flow presents near impellers, in industry the injection point is usually placed in that region. Consistent with this empirical rule, this simulation suggests that feeding at the corner of the reactor significantly hindered the progress of reaction.

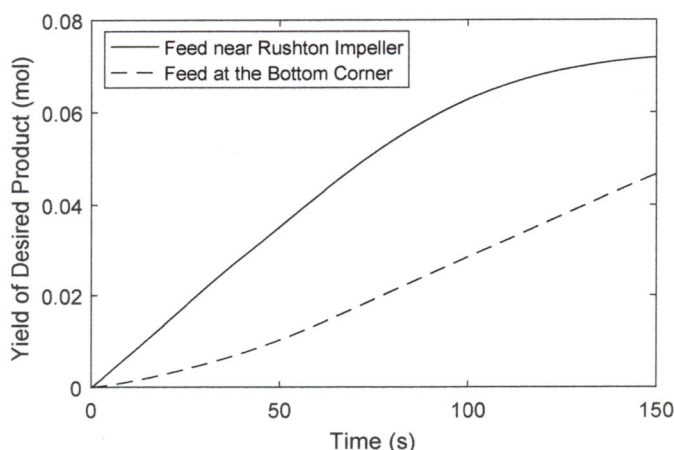

Figure 4. Simulated yield of the desired product when addition location is at the bottom corner and near Rushton impeller.

The location for feeding is then numerically optimized with the proposed compartmental model. As suggested in Table 1, it was found that irrespective of the number of stages, the maximum productivity is reached when reagents are injected at the tip of Rushton impeller, which suggests a higher mixing efficiency.

Table 1. Comparison of optimal feeding location for (a) constant rate feeding (b) two-stage dynamic feeding policy and (c) three-stage feeding policy.

Reagent Injection Policy	Optimal Injection Location	
	Height (m)	Radial Position (m)
Constant Rate Feeding	0.1–0.13	0.22–0.25
Two-stage Dynamic Feeding	0.1–0.13	0.22–0.25
Three-stage Dynamic Feeding	0.1–0.13	0.22–0.25

3.5.2. Optimal Rate of Feeding

The optimal feeding rate profiles determined for different reagent injection policies are illustrated in Figure 5. It can be concluded that the proposed methodology favors a decreasing feeding rate profile, which leads to an increased process productivity. The reason behind this productivity boost is studied through process dynamics. As illustrated in Table 2, approximately 4% increase in process productivity is achieved by implementing a dynamic operation.

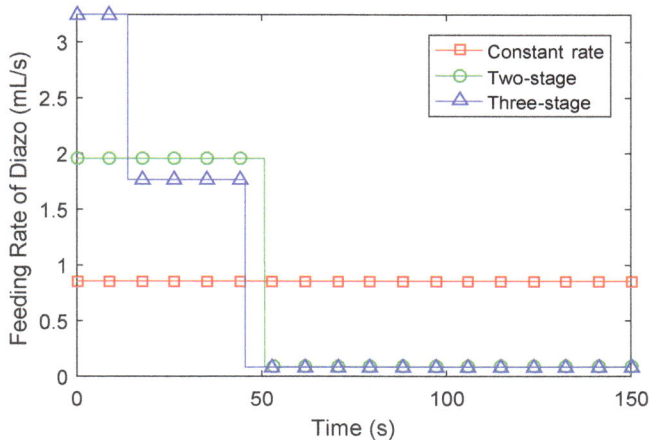

Figure 5. Optimal feeding rate profile solved for constant rate, two-stage and three-stage reagent injection policies.

Table 2. Comparison of optimal process productivity when (a) constant rate feeding (b) two-stage dynamic feeding policy and (c) three-stage feeding policy are adopted.

Reagent Injection Policy	Optimal Process Productivity ($)
Constant Rate Feeding	6162.90
Two-stage Dynamic Feeding	6410.43
Three-stage Dynamic Feeding	6411.76

Simulated trajectories of chemical species when different injection policies are employed is illustrated in Figure 6. It is suggested that through employing dynamic policies, the yield of desired product is not significantly enhanced (Figure 6c), which can be explain by the way we formulate this problem. Since the price of the desired product is 10 times as high as the price of diazonium, through the effort to maximize the overall profit, sufficient diazonium is fed to exhaust pyrazolone, which lead to similar yield of the product.

Nevertheless, dynamic feeding rate can improve process productivity through reducing the waste of raw material. As suggested in Figure 6a, considerable amount of diazonium is wasted if constant rate feeding policy is adopted. When reactant is fed at a constant rate, with the consumption of pyrazolone, diazonium would inevitably accumulate faster, which lead to material waste that compromises economic performance. By adjusting feeding rate as pyrazolone is deleted, maximum process profit can be achieved.

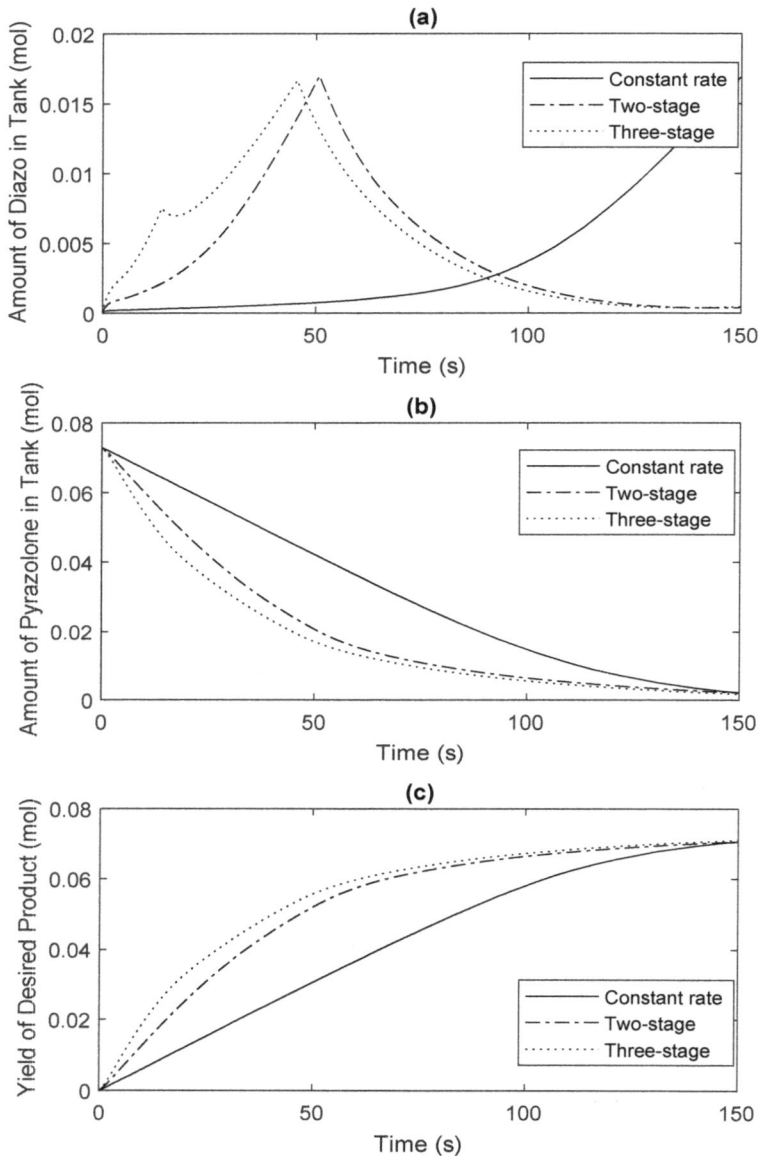

Figure 6. Simulated trajectories of (**a**) Diazo (**b**) Pyrazolone and (**c**) Desired product when optimal feeding policies solved with different number of stages are employed.

3.5.3. Traditional Process Design with Perfect-Mixing Assumptions

To illustrate the improvement of reaction selectivity by implementing this proposed methodology, perfect-mixing model is studied and compared with compartmental model. The same optimization algorithm is applied to the process dynamics model developed under perfect mixing assumption. Specifically, in this section, three-stage dynamic feeding rate is considered. Table 3 shows the simulated process productivity when different methodologies are employed. It can be concluded that by capturing the heterogeneity with CFD-based compartmental model, significant improvement to process productivity could be achieved. The reason behind this productivity boost is studied through process dynamics, as shown in Figures 7 and 8.

Table 3. Comparison between optimal operating conditions solved with perfect mixing model and the proposed methodology.

Methodology	Simulated Process Productivity ($)
Perfect-mixing Model	6162.90
CFD-based Compartmental Model	6410.43

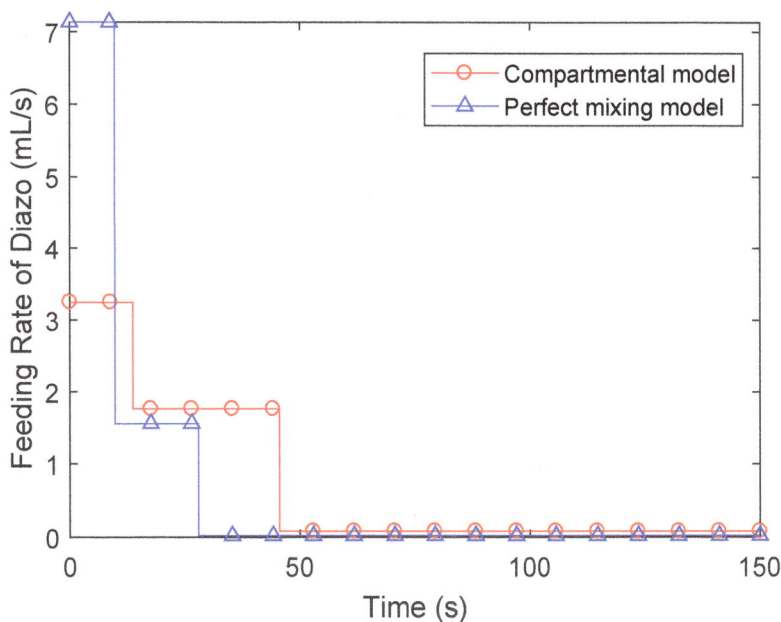

Figure 7. Optimal feeding rate profile solved with CFD-based compartmental model and perfect mixing model.

Optimal feeding rate profiles determined with different methodologies are illustrated in Figure 7. It is suggested that process design based on perfect-mixing assumption would lead to faster feeding at earlier period of process. Therefore, a high quantity of diazonium is accumulated in the earlier stage of process (Figure 8a). As a result, the undesired decomposition of diazonium is accelerated, which compromise reaction selectivity (Figure 8d). Moreover, without considering the insufficient consumption of pyrazolone due to imperfect mixing, the overall yield of desired product is hindered, which further reduced product productivity.

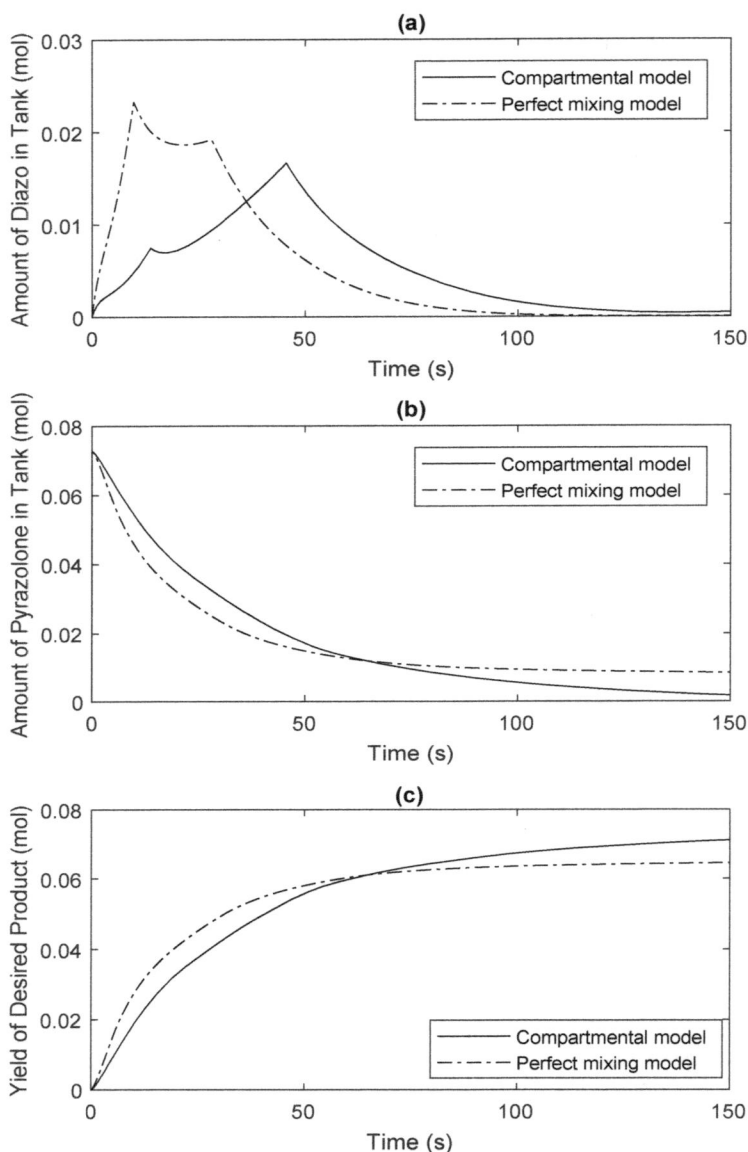

Figure 8. Simulated trajectories of (**a**) Diazonium (**b**) Pyrazolone (**c**) Desired product when different methodologies are employed.

4. Discussion

Mixing in turbulent flow can significantly influence reaction selectivity, therefore systematic analysis of mass transfer inside reactors is crucial for process design. Despite the increasing computational power available for gaining understanding of the mixing process, in-silico process design can still be inefficient in time and cost. In this proposed framework, by replacing repetitive dynamic CFD simulation with compartmental model, process design can be conducted in a timely and economically more efficient manner. Moreover, in this work surrogate-based optimization is implemented to numerically optimize process productivity. Comparing to genetic algorithm, which is most widely adopted in engineering design, surrogate–based method requires less simulations to find the optimal process design. As a result, the overall time efficiency can be significantly improved.

Rößger and Richter [25] have thoroughly compared state of the art CFD-based optimization methods based on a 2-parameter design problem. For that specific problem it has been demonstrated that direct optimization based on CFD simulation would require 1000 CFD simulations, which is computationally prohibitive. Hybrid methods based on data-driven models can reduce the required number of CFD simulations. However, as have been discussed, a large number of simulations are required to sufficiently train the data-driven model and prevent over-fitting. The reported computational time is 6.5 h running parallel on a 20-core Intel Xeon E5 V2 2.8 GHz processor.

In this manuscript, to optimize the process operation, 8 decision variables are considered. By implementing an efficient iterative SBO algorithm, 150 iterations are sufficient for the algorithm to converge to a good solution. Moreover, since each iteration is based on the compartmental model, which only require a single CFD simulation to construct. As presented in Table 4, single simulations were performed to determine the computational effort for one design evaluation. The overall computational time is 3 h, running on a single-core of an Intel Xeon E5 V3 workstation. It can be concluded that by implementing CFD-based compartmental modeling, significant time saving could be achieved if same hardware system is applied. In active pharmaceutical ingredient (API) plant design where a large number of simulations are conducted, computational expense could be reduced from weeks to h through the implementation of the proposed methodology. Moreover, considering that CFD is not freely available and requires special know-how, by cutting back dynamics CFD simulations, implementing compartmental modeling is economically beneficial. This approach has shown a good potential to characterize mixing in all the reactors in an API plant.

Table 4. Comparing computational expense between the proposed methodology and direct CFD simulation.

Methodology	Computational Expense for One Simulation (s)
Dynamic CFD simulation	10^4
CFD-based Compartmental Model	70

Despite the reduced model complexity, in this work the grid density is determined so that inhomogeneity inside reactor is sufficiently captured. Comparing to traditional process design strategies which are based on perfect-mixing assumption, better reaction selectivity could be achieved through compartmental model. As has been discussed in sub Section 3.5.3, by replacing traditional process design strategy with this proposed compartmental model, more than 10% increase in process profit has been achieved (Table 5).

Table 5. Comparison between optimal operating conditions solved with perfect mixing model and the proposed methodology.

Methodology	Simulated Process Productivity ($)
Perfect-mixing Model	5713.18
Compartmental Model (constant rate)	6126.90
Compartmental Model (dynamic rate)	6411.76

Furthermore, this proposed framework allows the design of dynamic operations. As illustrated in sub Section 3.5.2, by adapting feeding rate in time to account for the depletion of raw materials, material waste can be substantially reduced, which in turn leads to improved process productivity. Considering that for complex reaction networks commonly encountered in organic synthesis, process dynamics could be very complicated. By enabling the design of dynamic operations to fit the evolving chemical processes, this proposed framework has exhibited a great potential of productivity improvement.

5. Conclusions

CFD is a powerful tool to study mixing in chemical processes, and as such it has been applied widely for numerical optimization of reactor designs. Nevertheless, implementation of CFD to improve mixing through optimization of process operation is limited. A key reason is the computational complexity of CFD models. Even though some data-driven models are used to replace CFD to reduce time expense, to build confidence in the data-driven models requires large amount of CFD simulation. The implementation of meta-heuristic algorithms further increased the computational expense.

This paper has shown possible ways to address the inherent difficulty with two improvements. First, instead of using data-driven models to represent CFD simulation, compartmental model is implemented. Since compartmental model are built based on first-principle mass balance equations, it requires significantly less CFD simulations to drive, making it a better compromise between computational complexity and model accuracy. Second, GA is replaced with surrogate-based optimization based on RBF, which has been proven to be more efficient than GA. It should be noted that this surrogate-based optimization algorithm does not give global optimality guarantees, same as GA. The goals should be to find "good" or near optimal solutions when enough resources are provided.

The surrogate-based compartmental model optimization presented a compact and efficient structure, which can be easily implemented in other scenarios. Since in compartmental model space is discretized, which leads to mixed-integer nonlinear programing problems, surrogate-based optimization is an efficient way to address it. Moreover, compartmental model is built based on steady-state hydrodynamic solution from CFD, which is independent from models inside each compartment. Therefore, changing models in compartments does not require extra CFD simulations.

The proposed methodology allows for dynamic process operation design, which has shown a great potential of productivity improvement. Moreover, dynamic optimization of process operation improves process flexibility and agility to adjust to a more dynamic market. Finally, the developed simplification of complex transport model has the potential for advanced process monitoring and control for reactors with limited instrumentation, such as single-use bioreactors and fermenters.

It should be noted that, to leverage the result of CFD simulation, compartmental models are constructed based on steady-state hydrodynamic solution. However, this simplification is based on the assumption that the flow field is independent from chemical reaction. This assumption would limit the application of the proposed method in reactive flows. In addition, experimental calibration of CFD simulation is necessary build confidence in the predictive ability of the proposed methodology.

Author Contributions: Under supervision of M.I. and S.K., S.Y. was responsible for development of research methodology and optimization framework, model development, optimization and analysis of the results. P.F. contributed in model development.

Funding: This research received no external funding.

Acknowledgments: The authors gratefully acknowledge financial support from Gilead Science Inc. We also appreciate the discussions and technical inputs from Chiajen Lai of Gilead Science Inc.

Conflicts of Interest: The authors declare no conflicts of interest.

Abbreviations

CPP	Critical process parameters
CFD	Computational fluid dynamics
QbD	Quality by design
RTD	Residence time distribution
GA	Genetic algorithm
RBF	Radial basis function
CMC	Chemistry, manufacturing, and controls
ANN	Artificial neural network
RANS	Reynolds averaged Navier-Stocks

MRF	Multi-reference frame
API	Active pharmaceutical ingredient
SBO	Surrogate-based opitmization

List of Symbols:

c	Concentration
N	Mass flux
D	Diffusivity
R	Source
V	Volume
S	Surface area
Q	Mass flow rate
Da	Damköhler number
Pe	Péclet number
k	Reaction rate constant
L	Characteristic length
u	Characteristic velocity
P	Price
y	Yield
N_s	Number of feeding stages
N_c	Number of compartments
t	Duration
f	Feed rate

Subscripts:

i	ith species
j	jth compartment
m	mth stage of operation

References

1. Plutschack, M.B.; Pieber, B.; Gilmore, K.; Seeberger, P.H. The Hitchhiker's Guide to Flow Chemistry(II). *Chem. Rev.* **2017**, *117*, 11796–11893. [CrossRef]
2. Gobert, S.R.L.; Kuhn, S.; Braeken, L.; Thomassen, L.C.J. Characterization of Milli- and Microflow Reactors: Mixing Efficiency and Residence Time Distribution. *Org. Process. Res. Dev.* **2017**, *21*, 531–542. [CrossRef]
3. Fogler, H.S. *Essentials of Chemical Reaction Engineering*; Pearson Education: Upper Saddle River, NJ, USA, 2010.
4. Gresch, M.; Brugger, R.; Meyer, A.; Gujer, W. Compartmental Models for Continuous Flow Reactors Derived from CFD Simulations. *Environ. Sci. Technol.* **2009**, *43*, 2381–2387. [CrossRef]
5. Nienow, A.W. On impeller circulation and mixing effectiveness in the turbulent flow regime. *Chem. Eng. Sci.* **1997**, *52*, 2557–2565. [CrossRef]
6. Rosseburg, A.; Fitschen, J.; Wutz, J.; Wucherpfennig, T.; Schluter, M. Hydrodynamic inhomogeneities in large scale stirred tanks–Influence on mixing time. *Chem. Eng. Sci.* **2018**, *188*, 208–220. [CrossRef]
7. Levesque, F.; Bogus, N.J.; Spencer, G.; Grigorov, P.; McMullen, J.P.; Thaisrivongs, D.A.; Davies, I.W.; Naber, J.R. Advancing Flow Chemistry Portability: A Simplified Approach to Scaling Up Flow Chemistry. *Org. Process. Res. Dev.* **2018**, *22*, 1015–1021. [CrossRef]
8. Aubin, J.; Ferrando, M.; Jiricny, V. Current methods for characterising mixing and flow in microchannels. *Chem. Eng. Sci.* **2010**, *65*, 2065–2093. [CrossRef]
9. Commenge, J.M.; Falk, L. Villermaux-Dushman protocol for experimental characterization of micromixers. *Chem. Eng. Process.* **2011**, *50*, 979–990. [CrossRef]
10. Reckamp, J.M.; Bindels, A.; Duffield, S.; Liu, Y.C.; Bradford, E.; Ricci, E.; Susanne, F.; Rutter, A. Mixing Performance Evaluation for Commercially Available Micromixers Using Villermaux-Dushman Reaction Scheme with the Interaction by Exchange with the Mean Model. *Org. Process. Res. Dev.* **2017**, *21*, 816–820. [CrossRef]
11. Cheng, D.; Feng, X.; Cheng, J.C.; Yang, C. Numerical simulation of macro-mixing in liquid-liquid stirred tanks. *Chem. Eng. Sci.* **2013**, *101*, 272–282. [CrossRef]

12. Liu, L.; Barigou, M. Experimentally Validated Computational Fluid Dynamics Simulations of Multicomponent Hydrodynamics and Phase Distribution in Agitated High Solid Fraction Binary Suspensions. *Ind. Eng. Chem. Res.* **2014**, *53*, 895–908. [CrossRef]

13. Reinecke, S.F.; Deutschmann, A.; Jobst, K.; Hampel, U. Macro-mixing characterisation of a stirred model fermenter of non-Newtonian liquid by flow following sensor particles and ERT. *Chem. Eng. Res. Des.* **2017**, *118*, 1–11. [CrossRef]

14. Warmeling, H.; Behr, A.; Vorholt, A.J. Jet loop reactors as a versatile reactor set up—Intensifying catalytic reactions: A review. *Chem. Eng. Sci.* **2016**, *149*, 229–248. [CrossRef]

15. Farzan, P.; Ierapetritou, M.G. Integrated modeling to capture the interaction of physiology and fluid dynamics in biopharmaceutical bioreactors. *Comput. Chem. Eng.* **2017**, *97*, 271–282. [CrossRef]

16. Foli, K.; Okabe, T.; Olhofer, M.; Jin, Y.C.; Sendhoff, B. Optimization of micro heat exchanger: CFD, analytical approach and multi-objective evolutionary algorithms. *Int. J. Heat Mass Transf.* **2006**, *49*, 1090–1099. [CrossRef]

17. Uebel, K.; Rößger, P.; Prüfert, U.; Richter, A.; Meyer, B. CFD-based multi-objective optimization of a quench reactor design. *Fuel Process. Technol.* **2016**, *149*, 290–304. [CrossRef]

18. Chen, M.; Wang, J.; Zhao, S.; Xu, C.; Feng, L. Optimization of Dual-Impeller Configurations in a Gas–Liquid Stirred Tank Based on Computational Fluid Dynamics and Multiobjective Evolutionary Algorithm. *Ind. Eng. Chem. Res.* **2016**, *55*, 9054–9063. [CrossRef]

19. Na, J.; Kshetrimayum, K.S.; Lee, U.; Han, C. Multi-objective optimization of microchannel reactor for Fischer-Tropsch synthesis using computational fluid dynamics and genetic algorithm. *Chem. Eng. J.* **2017**, *313*, 1521–1534. [CrossRef]

20. De-Sheng, Z.; Jian, C.; Wei-Dong, S.H.I.; Lei, S.H.I.; Lin-Lin, G. Optimization of hydrofoil for tidal current turbine based on particle swarm optimization and computational fluid dynamic ethod. *Thermal Sci.* **2016**, *20*, 907–912.

21. Sierra-Pallares, J.; del Valle, J.G.; Paniagua, J.M.; Garcia, J.; Mendez-Bueno, C.; Castro, F. Shape optimization of a long-tapered R134a ejector mixing chamber. *Energy* **2018**, *165*, 422–438. [CrossRef]

22. Brar, L.S.; Elsayed, K. Analysis and optimization of cyclone separators with eccentric vortex finders using large eddy simulation and artificial neural network. *Sep. Purif. Technol.* **2018**, *207*, 269–283. [CrossRef]

23. Jung, I.; Kshetrimayum, K.S.; Park, S.; Na, J.; Lee, Y.; An, J.; Park, S.; Lee, C.-J.; Han, C. Computational Fluid Dynamics Based Optimal Design of Guiding Channel Geometry in U-Type Coolant Layer Manifold of Large-Scale Microchannel Fischer–Tropsch Reactor. *Ind. Eng. Chem. Res.* **2016**, *55*, 505–515. [CrossRef]

24. Park, S.; Na, J.; Kim, M.; Lee, J.M. Multi-objective Bayesian optimization of chemical reactor design using computational fluid dynamics. *Comput. Chem. Eng.* **2018**, *119*, 25–37. [CrossRef]

25. Rößger, P.; Richter, A. Performance of different optimization concepts for reactive flow systems based on combined CFD and response surface methods. *Comput. Chem. Eng.* **2018**, *108*, 232–239. [CrossRef]

26. Abu-Mostafa, Y.S.; Magdon-Ismail, M.; Lin, H.-T. *Learning from Dat*; AMLBook: New York, NY, USA, 2012; Volume 4.

27. Kapsoulis, D.; Tsiakas, K.; Trompoukis, X.; Asouti, V.; Giannakoglou, K. A PCA-assisted hybrid algorithm combining EAs and adjoint methods for CFD-based optimization. *Appl. Soft. Comput.* **2018**, *73*, 520–529. [CrossRef]

28. Muller, J.; Shoemaker, C.A.; Piche, R. SO-MI: A surrogate model algorithm for computationally expensive nonlinear mixed-integer black-box global optimization problems. *Comput. Oper. Res.* **2013**, *40*, 1383–1400. [CrossRef]

29. Boltersdorf, U.; Deerberg, G.; SCHLÜTER, S. Computational study of the effects of process parameters on the product distribution for mixing sensitive reactions and on distribution of gas in stirred tank reactors. *Recent Res. Dev. Chem. Eng.* **2000**, *4*, 15–43.

30. Guha, D.; Dudukovic, M.P.; Ramachandran, P.A.; Mehta, S.; Alvare, J. CFD-based compartmental modeling of single phase stirred-tank reactors. *AIChE J.* **2006**, *52*, 1836–1846. [CrossRef]

31. Nørregaard, A.; Bach, C.; Krühne, U.; Borgbjerg, U.; Gernaey, K.V. Hypothesis-driven compartment model for stirred bioreactors utilizing computational fluid dynamics and multiple pH sensors. *Chem. Eng. J.* **2019**, *356*, 161–169. [CrossRef]

32. Zhao, W.; Buffo, A.; Alopaeus, V.; Han, B.; Louhi-Kultanen, M. Application of the compartmental model to the gas-liquid precipitation of CO_2-$Ca(OH)_2$ aqueous system in a stirred tank. *AIChE J.* **2017**, *63*, 378–386. [CrossRef]

33. Bezzo, F.; Macchietto, S.; Pantelides, C.C. A general methodology for hybrid multizonal/CFD models. *Comput. Chem. Eng.* **2004**, *28*, 501–511. [CrossRef]

34. Vrabel, P.; van der Lans, R.; Cui, Y.Q.; Luyben, K. Compartment model approach: Mixing in large scale aerated reactors with multiple impellers. *Chem. Eng. Res. Des.* **1999**, *77*, 291–302. [CrossRef]

35. Du, J.; Johansen, T.A. Integrated Multilinear Model Predictive Control of Nonlinear Systems Based on Gap Metric. *Ind. Eng. Chem. Res.* **2015**, *54*, 6002–6011. [CrossRef]

36. Srilatha, C.; Morab, V.V.; Mundada, T.P.; Patwardhan, A.W. Relation between hydrodynamics and drop size distributions in pump–mix mixer. *Chem. Eng. Sci.* **2010**, *65*, 3409–3426. [CrossRef]

37. Bezzo, F.; Macchietto, S. A general methodology for hybrid multizonal/CFD models—Part II. *Automatic zoning. Comput. Chem. Eng.* **2004**, *28*, 513–525. [CrossRef]

38. Nienow, A.W.; Drain, S.M.; Boyes, A.P.; Mann, R.; El-Hamouz, A.M.; Carpenter, K.J. A new pair of reactions to characterize imperfect macro-mixing and partial segregation in a stirred semi-batch reactor. *Chem. Eng. Sci.* **1992**, *47*, 2825–2830. [CrossRef]

39. Conn, A.R.; Scheinberg, K.; Vicente, L.N. *Introduction to Derivative-Free Optimization*; Siam: Philadelphia, PA, USA, 2009.

40. Boukouvala, F.; Misener, R.; Floudas, C.A. Global optimization advances in Mixed-Integer Nonlinear Programming, MINLP, and Constrained Derivative-Free Optimization, CDFO. *Eur. J. Oper. Res.* **2016**, *252*, 701–727. [CrossRef]

41. Bhosekar, A.; Ierapetritou, M. Advances in surrogate based modeling, feasibility analysis, and optimization: A review. *Comput. Chem. Eng.* **2018**, *108*, 250–267. [CrossRef]

42. Regis, R.G.; Shoemaker, C.A. Improved strategies for radial basis function methods for global optimization. *J. Glob. Optim.* **2007**, *37*, 113–135. [CrossRef]

43. Oeuvray, R.; Bierlaire, M. BOOSTERS: A derivative-free algorithm based on radial basis functions. *Int. J. Model. Simul.* **2009**, *29*, 26–36. [CrossRef]

44. Wild, S.M.; Shoemaker, C.A. Global Convergence of Radial Basis Function Trust-Region Algorithms for Derivative-Free Optimization. *SIAM Rev.* **2013**, *55*, 349–371. [CrossRef]

45. Regis, R.G.; Wild, S.M. CONORBIT: constrained optimization by radial basis function interpolation in trust regions. *Optim. Meth. Softw.* **2017**, *32*, 552–580. [CrossRef]

46. Wang, Z.L.; Ierapetritou, M. A novel feasibility analysis method for black-box processes using a radial basis function adaptive sampling approach. *AIChE J.* **2017**, *63*, 532–550. [CrossRef]

47. Jones, D.R.; Schonlau, M.; Welch, W.J. Efficient global optimization of expensive black-box functions. *J. Glob. Optim.* **1998**, *13*, 455–492. [CrossRef]

48. Boukouvala, F.; Ierapetritou, M.G. Derivative-free optimization for expensive constrained problems using a novel expected improvement objective function. *AIChE J.* **2014**, *60*, 2462–2474. [CrossRef]

49. Regis, R.G. Trust regions in Kriging-based optimization with expected improvement. *Eng. Optim.* **2016**, *48*, 1037–1059. [CrossRef]

50. Beykal, B.; Boukouvala, F.; Floudas, C.A.; Sorek, N.; Zalavadia, H.; Gildin, E. Global optimization of grey-box computational systems using surrogate functions and application to highly constrained oil-field operations. *Comput. Chem. Eng.* **2018**, *114*, 99–110. [CrossRef]

51. Wang, Z.L.; Ierapetritou, M. Constrained optimization of black-box stochastic systems using a novel feasibility enhanced Kriging-based method. *Comput. Chem. Eng.* **2018**, *118*, 210–223. [CrossRef]

52. Fernandes, F.A.N. Optimization of Fischer-Tropsch synthesis using neural networks. *Chem. Eng. Technol.* **2006**, *29*, 449–453. [CrossRef]

53. Henao, C.A.; Maravelias, C.T. Surrogate-Based Superstructure Optimization Framework. *AIChE J.* **2011**, *57*, 1216–1232. [CrossRef]

54. Sen, O.; Gaul, N.J.; Choi, K.K.; Jacobs, G.; Udaykumar, H.S. Evaluation of multifidelity surrogate modeling techniques to construct closure laws for drag in shock-particle interactions. *J. Comput. Phys.* **2018**, *371*, 434–451. [CrossRef]

55. Müller, J. MISO: mixed-integer surrogate optimization framework. *Optim. Eng.* **2015**, *17*, 177–203. [CrossRef]

56. Zhang, W.P.; Wang, X.; Feng, X.; Yang, C.; Mao, Z.S. Investigation of Mixing Performance in Passive Micromixers. *Ind. Eng. Chem. Res.* **2016**, *55*, 10036–10043. [CrossRef]

57. Lin, X.Y.; Wang, K.; Zhang, J.S.; Luo, G.S. Liquid-liquid mixing enhancement rules by microbubbles in three typical micro-mixers. *Chem. Eng. Sci.* **2015**, *127*, 60–71. [CrossRef]

58. Fluent, A. *Ansys Fluent Theory Guide*; ANSYS Inc.: Canonsburg, PA, USA, 2011; Volume 15317, pp. 724–746.

processes

MDPI

Article

Adsorptive Properties of Poly(1-methylpyrrol-2-ylsquaraine) Particles for the Removal of Endocrine-Disrupting Chemicals from Aqueous Solutions: Batch and Fixed-Bed Column Studies

Augustine O. Ifelebuegu [1,*] , Habibath T. Salauh [1], Yihuai Zhang [1] and Daniel E. Lynch [2]

[1] School of Energy, Construction and Environment, Coventry University, Coventry CV1 5FB, UK;
     tittysalau@yahoo.com (H.T.S.); ifelechem@gmail.com (Y.Z.)
[2] Exilica Limited, The TechnoCentre, Coventry University Technology Park, Puma Way, Coventry CV1 2TT,
     UK; d.lynch@exilica.co.uk
*   Correspondence: A.Ifelebuegu@coventry.ac.uk; Tel.: +44-247-765-7690

Received: 27 July 2018; Accepted: 31 August 2018; Published: 4 September 2018

Abstract: The adsorptive properties of poly(1-methylpyrrol-2-ylsquaraine) (PMPS) particles were investigated in batch and column adsorption experiments as alternative adsorbent for the treatment of endocrine-disrupting chemicals in water. The PMPS particles were synthesised by condensing 3,4-dihydroxycyclobut-3-ene-1,2-dione (squaric acid) with 1-methylpyrrole in butanol. The results demonstrated that PMPS particles are effective in the removal of endocrine disrupting chemicals (EDCs) in water with adsorption being more favourable at an acidic pH, and a superior sorption capacity being achieved at pH 4. The results also showed that the removal of EDCs by the PMPS particles was a complex process involving multiple rate-limiting steps and physicochemical interactions between the EDCs and the particles. Gibbs free energy of −8.32 kJ/mole and −6.6 kJ/mol, and enthalpies of 68 kJ/mol and 43 kJ/mol, were achieved for the adsorption E2 and EE2 respectively The removal efficiencies of the EDCs by PMPS particles were comparable to those of activated carbon, and hence can be applied as an alternative adsorbent in water treatment applications.

Keywords: adsorption; PMPS particles; EDCs; breakthrough; fixed-bed column

1. Introduction

A variety of organic pollutants known as endocrine-disrupting chemicals (EDCs) are detected in both industrial and municipal wastewater, which have given rise to increased concerns about their existence in, and effects on the environment [1–3]. Multiple studies in literature have demonstrated that EDCs pose a significant health and environmental risk to humans and wildlife, due to their various sources and their ability to disrupt vital hormonal systems, even at nanogram levels [4–6]. Steroid hormones are among the most potent EDCs detected in wastewater effluents around the globe. The traditional wastewater treatment processes are usually not effective in the treatment of these EDCs to reduce them to potentially non-effect concentrations [5]. Adsorptions on to various adsorbents have been investigated as options for their removal from wastewater. Powdered and granular activated carbons have been considerably investigated as adsorbents for the treatment of EDCs in water and wastewater [7–10]. Other adsorbents that have been investigated include; carbon nanotubes [11–13], imprinted polymers [14], zeolites [15], and waste tealeaves [2].

Polysquaraines are a class of conjugated polymers that are synthesised by condensing 3,4-dihydroxycyclobut-3-ene-1,2-dione (or squaric acid) with a suitable electron-donating aromatic/heterocyclic ring system with at least two points for attachment to a squarate group [16]. Poly (1-methylpyrrol-2-ylsquaraine) (or PMPS) was one of three types of new squaraine compounds that were first reported in 1965 by Triebs and Jacob [17], but its unique particle shape was not reported until 2005 [18]. Extensive elemental metal adsorption studies on PMPS particles, along with physical studies on their internal porous nature, were published in 2016 [19]. They have also been recently reported as humidity sensing materials by Xiao et al., 2018 [20]. In the current study, we investigated the adsorptive properties of PMPS particles as an alternative adsorbent for the removal of steroid hormones in water and wastewater. The effects of experimental parameters on the performance of PMPS and the feasibility of using them in a continuous flow fixed column bed, were evaluated to establish the process parameters that are required for potential full-scale applications.

2. Material and Methods

2.1. Chemicals

The chemicals used in in this research including estradiol (E2), 17-ethiny estradiol (EE2), HPLC grade methanol and butanol, squaric acid, 1-methylpyrrole were of analytical grade. They were purchased from Sigma Aldrich (Saint-Quentin-Fallavier, UK). E2 and EE2 solutions were made in methanol. The PMPS particles were prepared at Exilica Limited, Coventry, UK.

2.2. PMPS Particles

PMPS particles were prepared as previously described [18,19]. Equimolar quantities of squaric acid and 1-methylpyrrole in butanol were refluxed for 18 h (Figure 1). The as-prepared PMPS particles were then filtered and washed with hot ethyl acetate for 4 h before drying at 60 °C. The physical properties of the as-prepared PMPS are presented in Table 1.

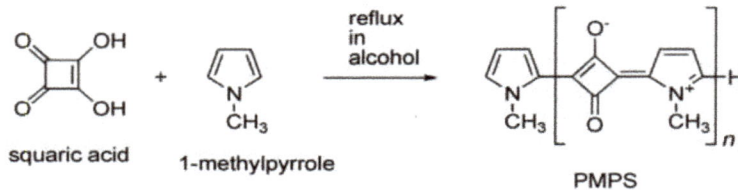

Figure 1. Synthesis of poly (1-methylpyrrol-2-ylsquaraine) (PMPS) particles.

Table 1. Physical characteristics of poly (1-methylpyrrol-2-ylsquaraine) PMPS particles [18,19].

Physical State	solid
Color	black
Specific gravity	1.3 to 1.5 g/cm^3
Average particle size	1.92 ± 0.05 μm [21]
Mean nominal rupture stress	493 ± 113 MPa
Mean deformation at rupture	+65% initial diameter
Surface area	~450 m^2/g

2.3. Batch Adsorption Experiments

Batch adsorption isotherm experiments using the bottle point method were carried out at varying concentrations of EDCs (0.1 to 2 mg/L) and temperatures (15 °C and 35 °C). Different weights of the PMPS particles were mixed with 200 mL of E2 and EE2 in 250 mL conical flasks. The experiments were conducted in duplicate, and the average values were reported. The pHs (2–5, 7, and 8) were varied using dilute HCl and NaOH solutions. The mixtures were stirred at a rate of 250 rpm using a magnetic stirrer for 120 min, with regular sampling and analyses at set intervals. The results obtained from the batch experiments were analysed using various kinetic (first and second order) and isotherm (Langmuir,

Freundlich, and Temkin) models. The thermodynamic parameters were also evaluated from kinetic and thermodynamic data. The amount of EDCs adsorbed was evaluated using Equation (1):

$$q_e(mg/g) \ = \frac{C_o - C_e(\frac{mg}{g})}{w(g)} v(L) \tag{1}$$

where C_o is the initial concentration of the adsorbate and C_e is the equilibrium concentration after adsorption, w is the dry weight of adsorbent (g), and v is the volume of aqueous solution (L).

2.4. Column Adsorption Experiment

Fixed bed column studies were conducted in a vertical down-flow glass column with internal diameter (d) of 2 cm and heights varying between 5 cm and 20 cm, with a feed reservoir of 10 L at the top of the column. The aqueous solutions of the EDCs were supplied through a peristaltic pump at a constant wetting rate. Treated samples were withdrawn at fixed interval and analysed for residual EDCs that are present. The Adam-Bohart, Thomas, and Yoon–Nelson models were used in predicting the breakthrough curve.

2.5. Analytical Procedures

E2 and EE2 were analysed using a Series 1050 HPLC system with a UV detector (HP) at a wavelength of 254 nm, column temperature of 25 °C, and a flow rate of 1.2 mL/min. The mobile phase was methanol (80%):water (20%) and the analysis was carried out in an isocratic mode. The detailed analytical protocol has been previously described [22,23].

3. Results and Discussion

3.1. Batch Adsorption Experiments

3.1.1. Influence of pH on Adsorptive Capacity of PMPS Particles

In the adsorption of adsorbates onto the surface of adsorbent materials, the pH of the solution plays an important role. This is due to the pH affecting the surface charge of adsorbents, and it can also trigger functional group dissociation on active sites [24]. The effects of pH on the adsorptive removal of E2 and EE2 were investigated at pH 2 to 8, and a temperature of 25 °C. The results are illustrated in Figure 2. It can be seen that the highest removal of the EDCs was achieved at pH 4. The adsorption capacity increased with pH, and then decreased after 4, showing decreased capacity at neutral to alkaline pH. With the pKa values of E2 and EE2 being 10.4 and 10.7 respectively [10]; at acidic pHs, they both exist predominantly in non-ionic molecular form and therefore are readily adsorbed onto the PMPS particles. However, as the pH increases, and beyond their dissociation constant (pKa), they become negatively charged, and hence the reduced sorption, due to electrostatic repulsion. Also, as pH increases, and the hydroxyl ion concentration increases, there is a trigger for the production of aqua-complexes [25,26], which could also contribute to the reduced adsorption capacity at higher pHs [10].

3.1.2. Effects of Contact Time and Adsorption Kinetics

The equilibrium times for the adsorption of the EDCs under consideration were determined at pH 4, for one hour. The results are shown in Figure 3. It can be seen that the adsorption of E2 and EE2 onto the PMPS particles was rapid within the first 10 min after which the rates slowed, attaining a steady state within the first 20 min, compared to the activated carbon, which has been reported to reach a steady state at 40 to 60 min for E2 and EE2 [2]. The results demonstrated that the removal rate of 93% was obtained within 20 min. This is comparable to the performance of activated carbon that was previously reported by Ifelebuegu [10]. Also, with a shorter time to achieve the steady state, PMPS may offer a smaller footprint in potential full-scale applications.

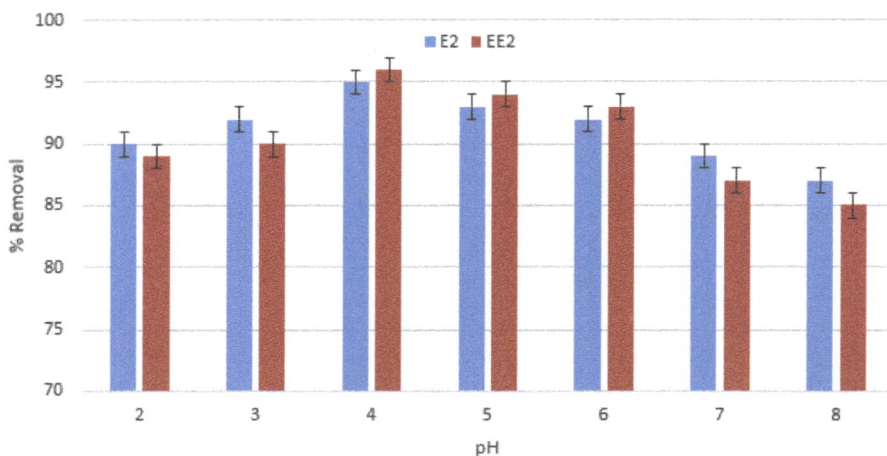

Figure 2. Effect of pH on the adsorptive removal of estradiol (E2) and 17-ethiny estradiol (EE2) by PMPS particles at a temperature of 25 °C (Error bars represent the standard deviation of the mean of the duplicates).

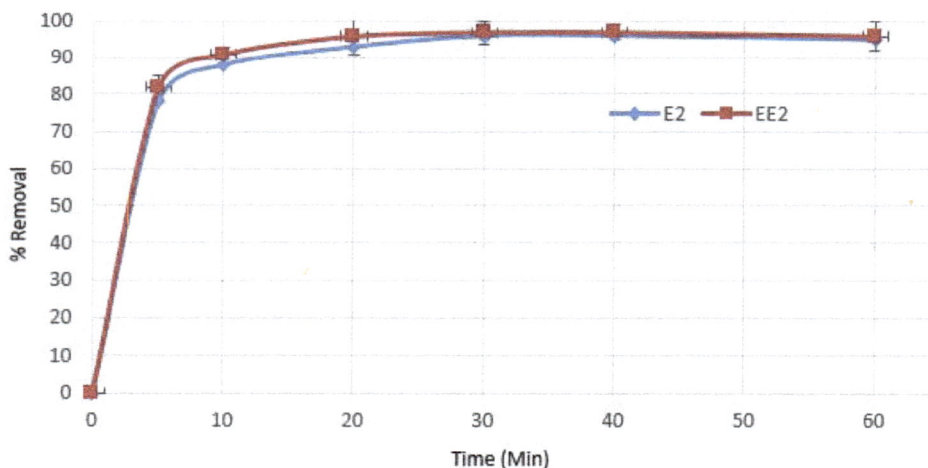

Figure 3. Removal of E2 and EE2 by PMPS particles at varying times and at a temperature of 25 °C (Error bars represent the standard deviation of the mean of the duplicates).

The kinetics of adsorption is a major consideration in the adsorption system design. It helps in predicting the rate-determining step and helps with obtaining the most optimum operating conditions during the design and optimization of full-scale plants [27]. In the current study, the pseudo-first-order and pseudo-second-order kinetic models were evaluated using the experimental data. The kinetic models for both pseudo-first order and pseudo-second order, as proposed by Lagergren [28] are expressed in Equations (2) and (3) respectively:

$$\frac{dq}{dt} = k_1 (q_e - q) \tag{2}$$

$$\frac{dq}{dt} = k_2 (q_e - q)^2 \tag{3}$$

where t is the contact time in min, k_1 is the first-order rate constant (min^{-1}), q_e is the amount of the adsorbate (mg/g) at equilibrium, and q is the amount of the adsorbate (mg/g) at time t. Equations (2) and (3) can also be expressed as Equations (4) and (5) respectively:

$$\ln C/C_o = -kt \tag{4}$$

$$\frac{1}{[C]} - \frac{1}{[C_o]} = kt \tag{5}$$

A plot of $\ln C/C_o$ against t gives a linear relationship from which k_1 is obtained from the slope of the graph. A plot of $1/[C] - 1/[C_o]$ against t will give a rate constant, k_2 $(Lmg^{-1}min^{-1})$, for the second-order kinetics. The results of the rate constants for E2 and EE2 are presented in Table 2.

The results of the kinetic experiments are presented in Table 2. It can be seen that the experimental data showed a better fit to the pseudo-second-order kinetic model with a higher correlation coefficient $(R^2 = 0.996$ for EE2 and 0.993 for E2), compared to the first-order kinetics which also showed a relatively good fit $(R^2 = 0.941$ for EE2 and 0.959 for E2). This suggests a complex interaction involving both physical and chemical interactions between the adsorbates and PMPS particles.

Table 2. Rate constants for the kinetic models.

Adsorbate	Pseudo-First-Order Kinetics			Pseudo-Second-Order Kinetics		
	Linear Equation	K	R^2	Linear Equation	K	R^2
EE2	0.016×-0.2366	0.017	0.941	0.0291×-0.072	0.029	0.996
E2	0.019×-0.2207	0.019	0.959	0.038×-0.2084	0.038	0.993

3.1.3. Adsorption Isotherm

An evaluation of the adsorption isotherms provided insight into how the adsorbate distributed between the solid and aqueous phases. It also helps to model and optimise the design parameters [5,29]. The Langmuir, Freundlich, and Temkin isotherm models were used in the evaluation of the experimental data from this work.

Langmuir Adsorption Isotherm

The Langmuir isotherm helps to describe the homogeneous adsorption where the adsorption sites exhibit equal affinities for the adsorbate without the movement of the molecules adsorbed in the plane of the adsorbent surface [30,31]. The linearised form of the Langmuir isotherm can be expressed as:

$$q_e = \frac{Q_m K_L C_e}{1 + K_L C_e} \tag{6}$$

where C_e (mg/L) is the equilibrium concentration of the EDCs,

q_e (mg/g) is the amount of PMPS adsorbed per unit mass,
Q_m (mg/g) is the maximum amount of PMPS per unit mass to form a monolayer on the surface,
K_L (L/Mg) is the isotherm constant related to the affinity of the binding site.

A plot of $1/q_e$ against $1/C_e$ gives a straight-line graph, with the values of Q_m and K_L being evaluated from the slope and intercept respectively.

A dimensionless constant or separation factor (R_L) is represented as:

$$R_L = \frac{1}{1 + K_L C_o} \tag{7}$$

R_L describes the favourable nature of the adsorption process; $R_L > 1$ is unfavourable, $R_L = 0$ is linear, $0 < R_L < 1$ is favourable, and $R_L = 0$ is irreversible [32].

Freundlich Adsorption Isotherm

The Freundlich adsorption isotherm model is an empirical correlation that describes the adsorption on heterogeneous surfaces [33]. The isotherm is expressed as:

$$Q_e = K_f C_e^{1/n} \quad n > 1 \tag{8}$$

The logarithmic form of the Freundlich model is expressed as:

$$LogQ_e = LogK_f + \frac{1}{n}LogC_e \tag{9}$$

where K_f is the Freundlich model constant,

n is a measure of the adsorption intensity,
C_e (mg/L) is the adsorbate concentration at equilibrium,
Q_e (mg/g) is the amount of PMPS adsorbed per unit mass,

The plot of $LogQ_e$ against $LogC_e$ gives a straight line and n and K_f was evaluated from the slope and intercept respectively.

Temkin Adsorption Isotherm

The Temkin models is expressed in Equation (10). It assumes that within the adsorption layers, the molecular heat of adsorption tends to decrease linearly with adsorbent coverage, due to the interaction between the adsorbate and the adsorbent [33]. The linearised form of the isotherm is described in Equations (11) and (13):

$$q_e = \frac{RT}{b} \ln(A_T C_e) \tag{10}$$

$$q_e = \frac{RT}{b}\ln A_T + \left[\frac{RT}{b}\right]\ln C_e \tag{11}$$

$$B = \frac{RT}{bT} \tag{12}$$

$$q_e = B \ln A_T + B \ln C_e \tag{13}$$

where A_T (L/g) is the Temkin binding constant (L/g), b is the Temkin isotherm constant, R is the universal gas constant (8.314 J/(mol K)), T is temperature at in Kelvin, and B (J/mol) is the constant related to heat of sorption. A_T and B were evaluated respectively from the intercept and slope of the plot of q_e against $\ln C_e$.

Table 3 shows the Langmuir, Freundlich, and Temkin coefficients for single solute adsorption isotherms. The results demonstrated that the Freundlich isotherm achieved the best fit compared to the Langmuir and Temkin isotherms, comparing the correlation coefficients. This suggests a multilayer adsorption. According to Kadirvelu et al., 2003 [34], a value of n > 1 represents a beneficial adsorption process indicating PMPS particles' affinity for steroid hormones. This also supports a more highly physical interaction between the adsorbates and PMPS. The Langmuir isotherm with relatively good fit also demonstrated that the adsorption of PMPS particles is favorable, with R_L values between 0 and 1. The adsorption of the adsorbates onto the adsorbents are generally very favorable when the value of R_L is between 0 and 1 [31,32].

Table 3. Isotherm model constants and correction coefficients.

Adsorbate	Langmuir			Freundlich			Temkin		
	q_m	K_L	R^2	K_f	n	R^2	A_T	B	R^2
E2	0.313	2.80	0.977	0.287	1.59	0.995	1.386	0.2518	0.975
EE2	0.327	2.53	0.979	0.285	1.618	0.996	1.393	0.2517	0.981

3.1.4. Adsorption Thermodynamics

The temperature effects on the sorption of the EDCs were evaluated at varying temperatures (15, 20, 25, 30, and 35 °C), with the other experimental variables being kept constant. There was an increase in the retention capacity with increasing temperature, suggesting a possible endothermic reaction and increased desolvation of E2 and EE2 molecules and diffusivity within the PMPS particles. The thermodynamic parameters of the adsorption process (Gibb's free energy (ΔG^0), enthalpy (ΔH^0), and entropy (ΔS^0)) were evaluated from Equations (14)–(16):

$$\left(\Delta G^0\right) = -RT\ln K_D \tag{14}$$

$$\left(\Delta G^0\right) = \left(\Delta H^0\right) - T\left(\Delta S^0\right) \tag{15}$$

$$-RT\ln K_D = \left(\Delta H^0\right) - T\left(\Delta S^0\right) \tag{16}$$

where R is universal gas constant, T (K) is the Kelvin temperature, and K_D (L/g) is the quantity of the EDCs that are adsorbed onto the PMPS. The plot of $\ln K_D$ vs. 1/T gave a straight line, with ΔH^0 and ΔS^0 values being evaluated from the slope and intercept of the graph, respectively. The results are reported in Table 4. Gibbs free energy values of up to −20 kJ/mol shows adsorptions that are controlled by electrostatic interactions between the PMPS particles and the EDCs [35,36]. The values obtained for both E2 and EE2 are less than 10 kJ/mol, indicating physical adsorption. However, the values of ΔH^0 obtained for E2 (68 kJ/mol) and EE2 (43 kJ/mol) suggest that the sorption process is chemical in nature. Ifelebuegu, 2011 [5] obtained values of 91 and 95 KJ/mol for the adsorptions of E2 and EE2 onto granular activated carbon. This supports the previous conclusions from the kinetic and isotherm models that the mechanism of sorption is a combination of physical and chemical interactions between the adsorbates and the PMPS particles. The positive entropy values of E2 = 0.26 J/mol K and EE2 = 0.17 J/mol K, demonstrate an increased randomness that enhanced the retention capacity of the PMPS particles with increasing temperature.

Table 4. Thermodynamic parameters for E2 and EE2 at pH 7.

Thermodynamic Parameter	E2	EE2
Gibbs Free Energy (kJ/mol)	−8.32	−6.6
Enthalpy (kJ/mol)	68	43
Entropy (J/mol K)	0.26	0.17

3.2. Column Studies

The column tests were carried out using both E2 and EE2. The results obtained were similar; consequently, only those of EE2 have been reported in this section. The breakthrough curve and operational parameters such as bed height, time to breakthrough, and feed flow rates are important parameters in a successful adsorption column design. The breakthrough of EE2 was evaluated at varying bed heights, adsorbate concentrations, and flow rates.

3.2.1. Effects of the Operating Parameters on the Breakthrough Curve

To evaluate the effect of flow rate, the experiments was conducted at varying flowrates of 2.5, 5.5, and 11.5 mL/min. Other experimental parameters were left constant (Z = 10 cm, C_o = 2 mg/L). The breakthrough curve is shown in Figure 4a. Initially, all the adsorbates were removed as the adsorption proceeded until breakthrough, and there was an abrupt increase in the effluent concentration of EE2. With the increase in flow rate, there was a decrease in the empty bed contact time (EBCT), while the height of the mass transfer zone increased. At a lower flow rate, EE2 had more contact time with the PMPS particles, hence the higher EBCT.

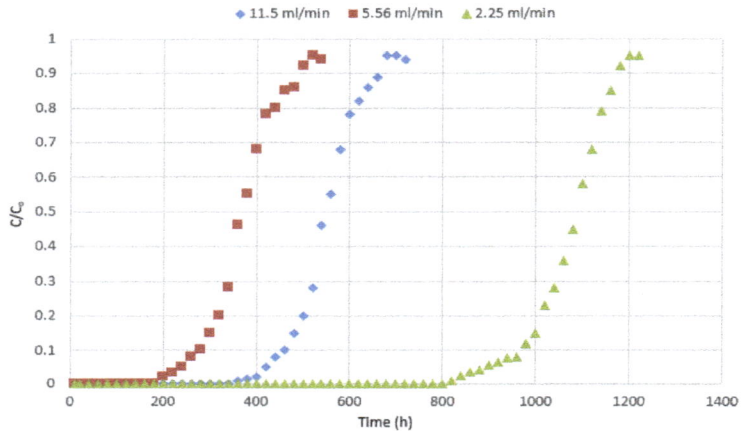

(a)

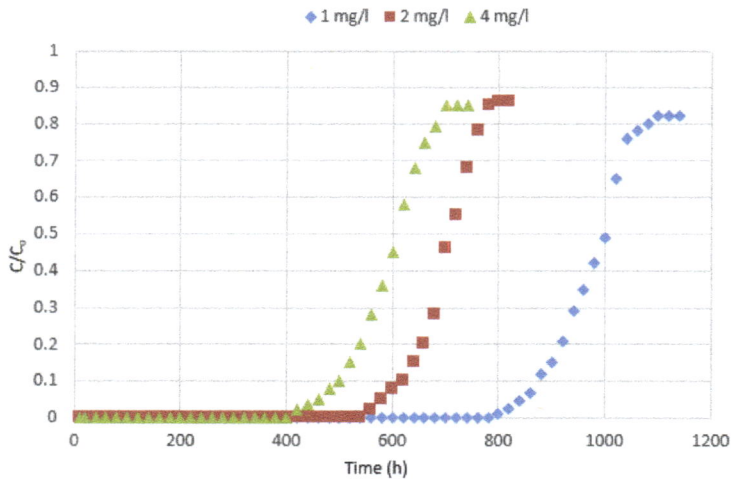

(b)

Figure 4. *Cont.*

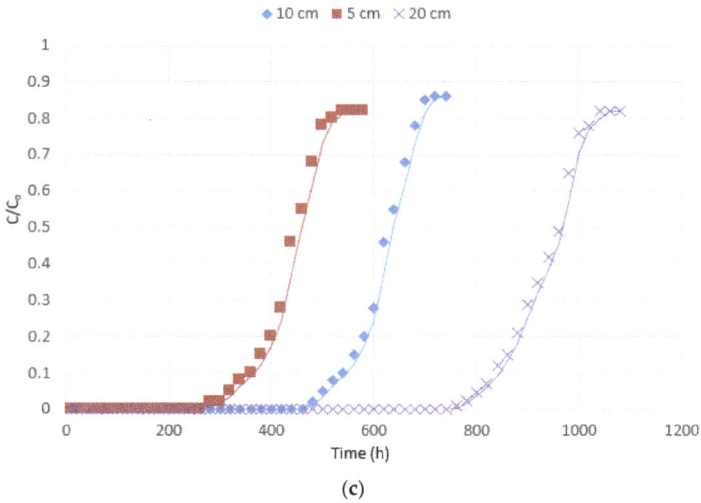

(c)

Figure 4. (a) Breakthrough curve for the adsorption of EE2 by PMPS at varying flow rates (Z = 10 cm, C₀ = 2 mg/L); (b) Breakthrough curve for the adsorption of EE2 by PMPS at varying influent concentrations (Z = 10 cm, flowrate = 5.5 mL/min); (c) Breakthrough curve for the adsorption of EE2 by PMPS at varying bed heights (C₀ = 2 mg/L, flowrate = 5.5 mL/min).

The effect of EE2 concentration was also investigate at varying concentrations of 1, 2, and 4 mg/L (Z = 10 cm, flowrate = 5.5 mL/min). The breakthrough curves (Figure 4b) demonstrated that with an increase in EE2 concentration, there was a corresponding increase in the steepness of the curves. This is attributed to the relative decrease in the mass transfer zone, as the influent EE2 concentration increased [37]. A higher influent concentration of EE2 made available a higher driving force that enabled the migration process to overcome the mass transfer resistance in the PMPS column, resulting in a higher adsorption of the EE2 in the column. Consequently, the higher exhaustion time demonstrated by the column with a lower EE2 concentration can be attributed to the lower driving force because of the reduced mass transfer resistance [38].

The effect of varying bed height was investigated at Z = 5, 10, and 20 cm. (C₀ = 2 mg/L, flowrate = 5.5 mL/min). The throughput volume of the treated solution (V_eff) increased as the bed height increased. The slope of the breakthrough curve was less steep with an increasing bed depth (Figure 4c), indicating the presence of an expanded mass transfer zone as the bed depth increased [39]. Yan et al., 2014 [40] reported similar results when the bed height was increased.

3.2.2. Breakthrough Modelling

The Adam–Bohart model is normally used for describing the front end of the breakthrough curve, and the model normally assumes that the rate of adsorption is proportional to the sorptive capacity of the adsorbent [41]. The Adam–Bohart expression is given by Equation (17):

$$\ln \frac{C_t}{C_o} = k_{AB}C_ot - k_{AB}N_o\frac{Z}{F} \tag{17}$$

where, C_o and C_t (mg/L) are the influent and effluent PMPS particle concentrations respectively. k_{AB} (L/mg min) is the Adam-Bohart constant, N_o (mg/L) is the saturation concentration, F (cm/min) is the superficial velocity, Z (cm) is the bed height, and t (min) is the total flow time. The values of N_o and k_{AB} were evaluated from the intercept and the slope of the linear plot of ln (C_t/C_o) against time (t) for all breakthrough curves. The values are shown in Table 5. It can be seen that k_{AB} values increased with increasing EE2 concentrations. However, there was no significant change with increasing bed height. Therefore, it can be inferred that the adsorption of EE2 onto PMPS particles involves several mechanisms, as there was no increase in the mass transfer resistance with increasing bed height.

Table 5. Adam–Bohart parameters at different conditions for the adsorption of EE2 on PMPS particles using a linear regression analysis.

C_o (mg L^{-1})	Z (cm)	v (mL min^{-1})	k_{AB} (L mg^{-1} min^{-1}) $\times 10^{-3}$	N_o (mg L^{-1})	R^2
1	10	5.56	6.70	239	0.955
2	10	5.56	7.55	275	0.940
4	10	5.56	6.75	377	0.899
2	5	5.56	7.75	182	0.946
2	20	5.56	7.55	364	0.927
2	10	2.25	7.00	170	0.988
2	10	11.50	8.0	761	0.974

The Adams-Bohart–plots for the removal of EE2 by PMPS particles at 10% and 50% breakthrough are shown in Figure 5. At 50% breakthrough, a plot of t against Z will produce a straight line that passes through the origin. It can be seen that the breakthrough curve did not pass through the origin, an indication that the adsorption of PMPS particles involves a complex process involving multiple rate-limiting steps [42]. This agrees with the finding in the batch adsorption tests.

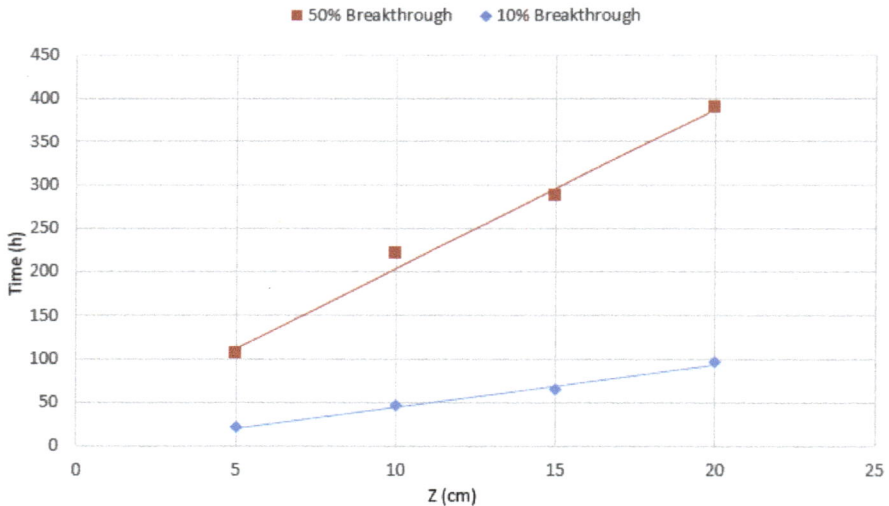

Figure 5. Adam-Bohart plot of EE2 removal by PMPS at 10% and 50% breakthrough points.

Thomas Model

The Thomas model was also used to estimate the adsorptive capacity and to predict the breakthrough curves. It uses the Langmuir isotherm and assumes the absence of axial dispersion with predominating film diffusion [43]. The Thomas Model can be expressed as Equation 18:

$$\ln\left(\frac{C_o}{C_t} - 1\right) = \frac{k_{TH}q_o m}{Q} - k_{TH}C_o t \tag{18}$$

where k_{TH} (mL/h/mg) is the Thomas kinetic constant, q_o (mg/g) is the maximum solid phase concentration, C_o (mg/L) is the influent EE2 concentration, Q (mL/h) is the volumetric flow rate, and t (h) is the total flow time. The values of k_{TH} and q_o were obtained from the plot of $\ln(C_o/C_t - 1)$ against time, using linear regression. The results are reported in Table 6. It is shown that the Thomas kinetic constant (k_{TH}) and the solid phase concentration of PMPS particles (q_o) increased with an increasing influent concentration of EE2. This indicated that the driving force for the adsorption of EE2 onto the PMPS was the concentration gradient [39].

Table 6. Thomas model parameters for EE2 adsorption onto PMPS particles.

C_o (mg/L)	Z (cm)	v (mL/min)	k_{TH} (mL/min·mg) $\times\ 10^{-3}$	q_o (mg/g)	R^2
1	10	5.56	7.50	1.67	0.956
2	10	5.56	8.70	1.86	0.966
4	10	5.56	8.85	2.50	0.930
2	5	5.56	8.25	1.31	0.956
2	20	5.56	7.70	1.39	0.930
2	10	2.25	7.40	1.11	0.987
2	10	11.50	6.6	2.56	0.923

The Yoon–Nelson Model

The Yoon and Nelson's model assumes that the rate of decrease in the probability of adsorption of each of the adsorbate molecules is directly proportional to the adsorbate molecule adsorption and the breakthrough on the adsorbent [44]. For a single-component system, the model is expressed as Equation (19):

$$\ln \frac{C_t}{C_o - C_t} = k_{YN}t - \tau k_{YN} \tag{19}$$

where k_{YN} (mL/min) is the Yoon-Nelson constant, and τ (min) is the time that is required for 50% breakthrough. k_{YN} and τ were evaluated from the plot of $\ln[C_t/(C_o - C_t)]$ against t using linear regression. The values obtained are presented in Table 7. The values of K_{YN} increased with increasing influent EE2 concentration and flow rate, while the time to achieve 50% breakthrough (τ) being decreased. This is attributed to the increase in the amount of the adsorbate available for adsorption, leading to a shorter breakthrough and exhaustion of the column.

Table 7. Yoon–Nelson parameters for the adsorption of EE2 on PMPS particles.

C_0 (mg/L)	Z (cm)	v (mL/min)	k_{YN} (mL/min)	τ (min)	R^2
1	10	5.56	0.0197	999	0.956
2	10	5.56	0.0224	589	0.992
4	10	5.56	0.0259	409	0.998
2	5	5.56	0.0226	456	0.989
2	20	5.56	0.0206	954	0.983
2	10	2.25	0.0201	1100	0.989
2	10	11.50	0.0213	411	0.993

4. Conclusions

The use of poly(1-methylpyrrol-2-ylsquaraine) particles as an alternative adsorbent for the removal of EDCs from water was investigated using batch and column adsorption experiments. The results of the experiments demonstrated that PMPS particles are effective in the removal of EDCs with the efficiencies of the removal of E2 and EE2 being comparable to those of granular activated carbon. The removal efficiencies were highest at an acidic pH, with the maximum sorption capacity being achieved at pH 4. The experimental data from both the batch and column adsorption tests showed that the removal mechanism is a complex process involving multiple rate limiting steps and physicochemical interactions. A Gibbs free energy of -8.32 kJ/mole and -6.6 kJ/mol, and enthalpies of 68 kJ/mol and 43 kJ/mol, were achieved for E2 and EE2 respectively, demonstrating the potential feasibility of deploying PMPS particles as an alternative adsorbent in a full-scale column bed for the removal of EDCs in water and wastewater treatment applications.

Author Contributions: A.O.I. conceived and designed the experiments and prepared the manuscript, D.E.L. synthesized the adsorbent, A.O.I., H.T.S. and Y.Z. carried out the laboratory experiments.

Funding: This research received no external funding

Acknowledgments: The authors thank Exilica Limited for their support in the synthesis of the PMPS adsorbent.

Conflicts of Interest: The authors declare no conflicts of interest

Nomenclature

A	Surface area (m^2)
A$_T$	Tempkin isotherm equilibrium binding constant (L/g)
EBCT	Empty bed contact time (h)
EDCs	Endocrine-disruptive compounds
E2	estradiol
EE2	17α-ethynylestradiol
C$_e$	E2 and EE2 concentration at equilibrium (mg/L)
C$_o$	E2 and EE2 concentration at time '0' (mg/L)
k$_1$	Pseudo first-order rate constants (min^{-1})
k$_2$	Pseudo-second-order rate constants (g/mg min)
k$_L$	Langmuir's sorption isotherm constant (L/mg)
k$_f$	Sorption capacity derived from Freundlich's isotherm model (L/mg)
k$_{TH}$	Thomas kinetic constant (mL/h/mg)
Q	Flow rate (mL/min)
Q$_m$	Sorption capacity of PMPS from the Langmuir's isotherm (mg/g)
q$_e$	Solid phase concentration of EE2 at contact time (t, min)(mg/g)
MTZ	Mass transfer zone to length of the column ratio (m)
t	Contact time (min)
t$_b$	Time to breakthrough (h)
t$_e$	Time to exhaustion (h)
Z	Total bed depth (cm)
1/n	Sorption intensity derived from Freundlich's isotherm model
ΔG^o	Change in free energy (kJ/mol)
ΔH^o	Change in enthalpy (kJ/mol)
ΔS^o	Change in entropy (J/kmol)

References

1. Nakada, N.; Tanishima, T.; Shinohara, H.; Kiri, K.; Takada, H. Pharmaceutical chemicals and endocrine disrupters in municipal wastewater in Tokyo and their removal during activated sludge treatment. *Water Res.* **2006**, *40*, 3297–3303. [CrossRef] [PubMed]

2. Ifelebuegu, A.O.; Ukpebor, J.E.; Obidiegwu, C.C.; Kwofi, B.C. Comparative potential of black tea leaves waste to granular activated carbon in adsorption of endocrine disrupting compounds from aqueous solution. *Glob. J. Environ. Sci. Manag.* **2015**, *1*, 205–214.

3. Ifelebuegu, A.O.; Ukpebor, J.; Nzeribe-Nwedo, B. Mechanistic evaluation and reaction pathway of UV photo-assisted Fenton-like degradation of progesterone in water and wastewater. *Int. J. Environ. Sci. Technol.* **2016**, *13*, 2757–2766. [CrossRef]

4. Dalrymple, O.K.; Yeh, D.H.; Trotz, M.A. Removing pharmaceuticals and endocrine-disrupting compounds from wastewater by photocatalysis. *J. Chem. Technol. Biotechnol.* **2007**, *82*, 121–134. [CrossRef]

5. Ifelebuegu, A.O. The fate and behaviour of selected endocrine disrupting chemicals in full scale wastewater and sludge treatment unit processes. *Int. J. Environ. Sci. Technol.* **2011**, *8*, 245–254. [CrossRef]

6.   Kortenkamp, A. Endocrine disruptors: The burden of endocrine-disrupting chemicals in the USA. *Nat. Rev. Endocrinol.* **2017**, *13*, 6–7. [CrossRef] [PubMed]

7.   Choi, K.J.; Kim, S.G.; Kim, C.W.; Kim, S.H. Effects of activated carbon types and service life on removal of endocrine disrupting chemicals: Amitrol, nonylphenol, and bisphenol-A. *Chemosphere* **2005**, *58*, 1535–1545. [CrossRef] [PubMed]

8.   Ifelebuegu, A.O.; Lester, J.N.; Churchley, J.; Cartmell, E. Removal of an endocrine disrupting chemical (17α-ethinyloestradiol) from wastewater effluent by activated carbon adsorption: Effects of activated carbon type and competitive adsorption. *Environ. Technol.* **2006**, *27*, 1343–1349. [CrossRef] [PubMed]

9.   Kumar, A.K.; Mohan, S.V. Endocrine disruptive synthetic estrogen (17α-ethynylestradiol) removal from aqueous phase through batch and column sorption studies: Mechanistic and kinetic analysis. *Desalination* **2011**, *276*, 66–74. [CrossRef]

10. Ifelebuegu, A.O. Removal of steriod hormones by activated carbon adsorption—Kinetic and thermodynamic studies. *J. Environ. Prot.* **2012**, *3*, 469. [CrossRef]

11. Joseph, L.; Zaib, Q.; Khan, I.A.; Berge, N.D.; Park, Y.G.; Saleh, N.B.; Yoon, Y. Removal of bisphenol A and 17α-ethinyl estradiol from landfill leachate using single-walled carbon nanotubes. *Water Res.* **2011**, *45*, 4056–4068. [CrossRef] [PubMed]

12. Heo, J.; Flora, J.R.; Her, N.; Park, Y.G.; Cho, J.; Son, A.; Yoon, Y. Removal of bisphenol A and 17β-estradiol in single walled carbon nanotubes–ultrafiltration (SWNTs–UF) membrane systems. *Sep. Purif. Technol.* **2012**, *90*, 39–52. [CrossRef]

13. Zaib, Q.; Khan, I.A.; Saleh, N.B.; Flora, J.R.; Park, Y.G.; Yoon, Y. Removal of bisphenol A and 17β-estradiol by single-walled carbon nanotubes in aqueous solution: Adsorption and molecular modeling. *Water Air Soil Pollut.* **2012**, *223*, 3281–3293. [CrossRef]

14. Le Noir, M.; Plieva, F.; Hey, T.; Guieysse, B.; Mattiasson, B. Macroporous molecularly imprinted polymer/cryogel composite systems for the removal of endocrine disrupting trace contaminants. *J. Chromatogr. A* **2007**, *1154*, 158–164. [CrossRef] [PubMed]

15. Dong, Y.; Wu, D.; Chen, X.; Lin, Y. Adsorption of bisphenol A from water by surfactant-modified zeolite. *J. Colloid Interface Sci.* **2010**, *348*, 585–590. [CrossRef] [PubMed]

16. Lynch, D.E. Pyrrolyl Squaraines—Fifty Golden Years. *Metals* **2015**, *5*, 1349–1370. [CrossRef]

17. Triebs, A.; Jacob, K. Cyclotrimethine dyes derived from squaric acid. *Angew. Chem. Int. Ed.* **1965**, *4*, 694. [CrossRef]

18. Lynch, D.E.; Nawaz, Y.; Bostrom, T. Preparation of sub-micrometer silica shells using poly (1-methylpyrrol-2-ylsquaraine). *Langmuir* **2005**, *21*, 6572–6575. [CrossRef] [PubMed]

19. Lynch, D.E.; Bennett, J.B.; Bateman, M.J.; Reeves, C.R. The uptake of metal elements into poly(1-methylpyrrol-2-ylsquaraine) particles and a study of their porosity. *Adsorpt. Sci. Technol.* **2016**, *34*, 176–192. [CrossRef]

20. Xiao, X.; Zhang, Q.J.; He, J.H.; Xu, Q.F.; Li, H.; Li, N.J.; Chen, D.Y.; Lu, J.M. Polysquaraines: Novel humidity sensor materials with ultra-high sensitivity and good reversibility. *Sens. Actuators B Chem.* **2018**, *255*, 1147–1152. [CrossRef]

21. Begum, S.; Jones, I.P.; Jiao, C.; Lynch, D.E.; Preece, J.A. Characterisation of hollow Russian doll microspheres. *J. Mater. Sci.* **2010**, *45*, 3697–3706. [CrossRef]

22. Ifelebuegu, A.O.; Theophilus, S.C.; Bateman, M.J. Mechanistic evaluation of the sorption properties of endocrine disrupting chemicals in sewage sludge biomass. *Int. J. Environ. Sci. Technol.* **2010**, *7*, 617–622. [CrossRef]

23. Ifelebuegu, A.O.; Ezenwa, C.P. Removal of endocrine disrupting chemicals in wastewater treatment by Fenton-like oxidation. *Water Air Soil Pollut.* **2011**, *217*, 213–220. [CrossRef]

24. Fan, L.; Luo, C.; Sun, M.; Qiu, H.; Li, X. Synthesis of magnetic β-cyclodextrin–chitosan/graphene oxide as nanoadsorbent and its application in dye adsorption and removal. *Colloids Surf. B Biointerfaces* **2013**, *103*, 601–607. [CrossRef] [PubMed]

25. Mohan, S.V.; Shailaja, S.; Krishna, M.R.; Sarma, P.N. Adsorptive removal of phthalate ester (Di-ethyl phthalate) from aqueous phase by activated carbon: A kinetic study. *J. Hazard. Mater.* **2007**, *146*, 278–282. [CrossRef] [PubMed]

26. Kumar, A.K.; Mohan, S.V.; Sarma, P.N. Sorptive removal of endocrine-disruptive compound (estriol, E3) from aqueous phase by batch and column studies: Kinetic and mechanistic evaluation. *J. Hazard. Mater.* **2009**, *164*, 820–828. [CrossRef] [PubMed]

27. Kalavathy, M.H.; Karthikeyan, T.; Rajgopal, S.; Miranda, L.R. Kinetic and isotherm studies of Cu (II) adsorption onto H$_3$PO$_4$-activated rubber wood sawdust. *J. Colloid Interface Sci.* **2005**, *292*, 354–362. [CrossRef] [PubMed]

28. Langergren, S. About the theory of so-called adsorption of soluble substances. *Sven. Vetenskapsakad. Handlingnaar.* **1898**, *24*, 1–39.

29. Parab, H.; Joshi, S.; Shenoy, N.; Lali, A.; Sarma, U.S.; Sudersanan, M. Determination of kinetic and equilibrium parameters of the batch adsorption of Co (II), Cr (III) and Ni (II) onto coir pith. *Process Biochem.* **2006**, *41*, 609–615. [CrossRef]

30. Pérez-Marín, A.; Zapata, V.M.; Ortuno, J.; Aguilar, M.; Sáez, J.; Lloréns, M. Removal of cadmium from aqueous solutions by adsorption onto orange waste. *J. Hazard. Mater.* **2007**, *139*, 122–131. [CrossRef] [PubMed]

31. Foo, K.Y.; Hameed, B.H. Insights into the modeling of adsorption isotherm systems. *Chem. Eng. J.* **2010**, *156*, 2–10. [CrossRef]

32. Yuan, X.; Xia, W.; An, J.; Yin, J.; Zhou, X.; Yang, W. Kinetic and thermodynamic studies on the phosphate adsorption removal by dolomite mineral. *J. Chem.* **2015**, *2015*, 853105. [CrossRef]

33. Dada, A.; Olalekan, A.; Olatunya, A.; Dada, O. Langmuir, Freundlich, Temkin and Dubinin-radushkevich isotherms studies of equilibrium sorption of Zn2 Unto phosphoric acid modified rice husk. *J. Appl. Chem.* **2012**, *3*, 38–45.

34. Kadirvelu, K.; Kavipriya, M.; Karthika, C.; Radhika, M.; Vennilamani, N.; Pattabhi, S. Utilization of various agricultural wastes for activated carbon preparation and application for the removal of dyes and metal ions from aqueous solutions. *Bioresour. Technol.* **2003**, *87*, 129–132. [CrossRef]

35. Horsfall, M.; Spiff, A.I.; Abia, A.A. Studies on the influence of mercaptoacetic acid (MAA) modification of cassava (Manihot sculenta cranz). waste biomass on the adsorption of Cu^{2+} and Cd^{2+} from aqueous solution. *Bull. Korean Chem. Soc.* **2004**, *25*, 969–976.

36. Abdel Ghani, N.T.; Elchaghaby, G.A. Influence of operating conditions on the removal of Cu, Zn, Cd and Pb ion from wastewater by adsorption. *Int. J. Environ. Sci. Technol.* **2007**, *4*, 451–456. [CrossRef]

37. Woumfo, E.D.; Siéwé, J.M.; Njopwouo, D. A fixed-bed column for phosphate removal from aqueous solutions using an andosol-bagasse mixture. *J. Environ. Manag.* **2015**, *151*, 450–460. [CrossRef] [PubMed]

38. Rout, P.R.; Bhunia, P.; Dash, R.R. Modeling isotherms, kinetics and understanding the mechanism of phosphate adsorption onto a solid waste: ground burnt patties. *J. Environ. Chem. Eng.* **2014**, *2*, 1331–1342. [CrossRef]

39. Nur, T.; Loganathan, P.; Nguyen, T.; Vigneswaran, S.; Singh, G.; Kandasamy, J. Batch and column adsorption and desorption of fluoride using hydrous ferric oxide: Solution chemistry and modeling. *Chem. Eng. J.* **2014**, *247*, 93–102. [CrossRef]

40. Yan, Y.; Sun, X.; Ma, F.; Li, J.; Shen, J.; Han, W.; Liu, X.; Wang, L. Removal of phosphate from etching wastewater by calcined alkaline residue: Batch and column studies. *J. Taiwan Inst. Chem. Eng.* **2014**, *45*, 1709–1716. [CrossRef]

41. Goel, J.; Kadirvelu, K.; Rajagopal, C.; Garg, V.K. Removal of lead (II) by adsorption using treated granular activated carbon: Batch and column studies. *J. Hazard. Mater.* **2005**, *125*, 211–220. [CrossRef] [PubMed]

42. Zulfadhly, Z.; Mashitah, M.D.; Bhatia, S. Heavy metals removal in fixed-bed column by the macro fungus Pycnoporus sanguineus. *Environ. Pollut.* **2001**, *112*, 463–470. [CrossRef]

43. Ghasemi, M.; Keshtkar, A.R.; Dabbagh, R.; Safdari, S.J. Biosorption of uranium (VI) from aqueous solutions by Ca-pretreated Cystoseira indica alga.Breakthrough curves studies and modeling. *J. Hazard. Mater.* **2011**, *189*, 141–149. [CrossRef] [PubMed]

44. Nwabanne, J.; Igbokwe, P. Adsorption performance of packed bed column for the removal of lead (II) using oil palm fibre. *Int. J. Appl. Sci. Technol.* **2012**, *2*, 5.

![processes logo] *processes*

MDPI

Article

Input Shaping Predictive Functional Control for Different Types of Challenging Dynamics Processes

Muhammad Abdullah [1,2] and John Anthony Rossiter [1,*]

[1] Department of Automatic Control and System Engineering, University of Sheffield, Mappin Street, Sheffield S1 3JD, UK; MAbdullah2@sheffield.ac.uk

[2] Department of Mechanical Engineering, International Islamic University Malaysia, Jalan Gombak, Kuala Lumpur 53100, Malaysia

* Correspondence: j.a.rossiter@sheffield.ac.uk

Received: 2 July 2018; Accepted: 23 July 2018; Published: 7 August 2018

check for updates

Abstract: Predictive functional control (PFC) is a fast and effective controller that is widely used for processes with simple dynamics. This paper proposes some techniques for improving its reliability when applied to systems with more challenging dynamics, such as those with open-loop unstable poles, oscillatory modes, or integrating modes. One historical proposal considered is to eliminate or cancel the undesirable poles via input shaping of the predictions, but this approach is shown to sometimes result in relatively poor performance. Consequently, this paper proposes to shape these poles, rather than cancelling them, to further enhance the tuning, feasibility, and stability properties of PFC. The proposed modification is analysed and evaluated on several numerical examples and also a hardware application.

Keywords: predictive control; unstable; underdamped; integrating; input shaping

1. Introduction

Predictive functional control (PFC) is a simple controller that is effective for small-scale single-input-single-output (SISO) applications, especially for low-order and stable processes [1–3]. The main advantages of PFC compared to its prime competitor—that is, proportional integral derivative (PID)—are its ability to handle constraints and its transparent tuning parameters. Indeed, it must be emphasised that the performance of PFC should not be benchmarked against more advanced model predictive control (MPC) strategies [4], because the implementation is relatively much cheaper and requires only low computation with very straightforward coding [5,6].

Nevertheless, despite its apparently attractive benefits, PFC has received relatively little attention in the academic literature because of the lack of a priori stability guarantees [7,8], which are possible with more advanced MPC approaches. Consequently, several recent works have developed the basic concept of PFC to improve its overall tuning properties while providing a confident assurance in the resulting closed-loop performance and constraint handling [9–12]. However, most of these modifications perform well only with low-order and simple dynamical processes. For a system with open-loop unstable poles, significant underdamping, or integrating dynamics, PFC is quite challenging to tune effectively [13], and the resulting divergent or oscillating predictions may give rise to infeasibility and/or robustness issues.

PFC practitioners often deploy a type of cascade structure to handle a challenging dynamics process, where an inner loop is used to improve the dynamics for an outer loop to control [14,15]. This modification enables a user to retain an independent model (IM) structure that handles uncertainties, while retaining a similar tuning concept as the conventional approach. Nevertheless, the inner PFC can only accept a proportional-type controller to avoid any overcomplication when handling constraints [6]. This restriction makes it difficult to have a systematic selection of gain, rather than an ad hoc approach. Moreover, a back-calculation or anti-wind-up technique is required to avoid any saturation while satisfying the constraints [6], which also can be conservative and thus affect performance.

A fundamental conceptual weakness of a simplistic PFC approach is that one is basing decisions on an open-loop prediction, which may have undesirable, possibly divergent, dynamics; matching this to a desirable closed-loop dynamic will lead to an ill-posed approach. Hence, an alternative approach is to pre-stabilise/pre-shape the output predictions by shaping the future input dynamics so that the effect of unwanted poles on the predictions are alleviated or cancelled [16]. This modification can retain the systematic tuning concept of PFC, in addition to facilitating a recursive feasibility guarantee feature for constraint handling. However, the performance of this control law is not always desirable, since the cancellation of specified modes (poles) within the predictions using a minimal number of control input changes often requires an aggressive input trajectory [16–18], which is not implementable or ideal for some cases.

In practice, less aggressive shaping to remove the undesirable modes from the prediction [19] is preferable, and this forms the main thrust of the proposal in this paper. More specifically, the proposal here is that, rather than using a small, finite number of control moves (effectively, finite impulse response (FIR) parameterisation), such as in Generalized Predictive Control (GPC) Added the defination and conventional PFC, the future input moves will be parameterised using an infinite impulse response (IIR) instead. The preference for IIR over FIR is due to IIR being more convenient to manipulate and define than a high-order FIR, in general. In turn, by allowing the output modes to evolve over many more samples, the required input will be less aggressive. Nevertheless, due to the desire for simplicity and transparency that is a central tenet of PFC, in this paper, we choose not to generalise the parameterisation for different dynamics. Instead, this work seeks to provide some rigour on how to effectively and systematically shape the future input for a given dynamic and, moreover, how to ensure some recursive feasibility properties during constraint handling.

Section 2 of this paper presents a brief formulation of conventional PFC. Section 3 introduces the concept of input shaping PFC, together with the constraint handling approach. Section 4 demonstrate the proposed algorithms on several numerical examples and also on some laboratory hardware. Section 5 provides the conclusions.

2. Conventional PFC

This section presents, in brief, the main concepts, notation, and formulation of PFC, together with a systematic constraint handling approach. For more detailed derivations, theory, and concepts of PFC, an interested reader may refer to the references, e.g., [5,6,13,20]. Without loss of generality, a controlled autoregressive and integrated moving average (CARIMA)-based model is used instead of the standard independent model (IM) structure to derive the required unbiased predictions, as this form is more amenable to the algebra required to implement the shaping. Hence, the model will take the form:

$$a(z)y_k + b(z)u_k + \frac{\zeta_k}{\Delta(z)}; \quad \Delta(z) = 1 - z^{-1} \tag{1}$$

where $b(z) = b_1 z^{-1} + ...$, $a(z) = 1 + a_1 z^{-1} + ...$, y_k, u_k are the outputs and inputs, respectively, at sample k, and ζ_k is an unknown zero mean random variable used to capture uncertainty.

2.1. Control Law

PFC design is based on the assumption that a closed-loop response should follow a first-order dynamic from the current state y_k to the desired target r [20]. In practice, one aims to achieve this by ensuring a matching using the open-loop prediction, but only at a single point. In other words, the predicted output trajectory is chosen to satisfy the following equality:

$$y_{k+n|k} = (1 - \lambda^n)r + \lambda^n y_k \tag{2}$$

where $y_{k+n_y|k}$ is the n-step ahead system prediction at sample time k, and λ is the desired closed-loop pole that will determine the speed of convergence. PFC practitioners typically select the desired closed-loop time response (CLTR) which has an explicit link with the corresponding closed-loop pole, that is, $\lambda = e^{\frac{-3T}{CLTR}}$, where T is the sampling period [6].

The predictions for the CARIMA model (1) are standard in the literature (e.g., [4,21]), so only the final form is given here. For input increments Δu_{k+i}, the n-step ahead future output prediction is formed as:

$$\underset{\rightarrow}{y}_{k+1|k} = H\Delta \underset{\rightarrow}{u}_k + P\Delta \underset{\leftarrow}{u}_k + Q\underset{\leftarrow}{y}_k \tag{3}$$

where the left and right arrow underlying vectors represent past and future variables, respectively. The parameters H, P, Q depend on the model parameters, and for a model of order m:

$$\Delta \underset{\rightarrow}{u}_k = \begin{bmatrix} \Delta u_k \\ \Delta u_{k+1} \\ \vdots \\ \Delta u_{k+n-1} \end{bmatrix} ; \Delta \underset{\leftarrow}{u}_k = \begin{bmatrix} \Delta u_{k-1} \\ \Delta u_{k-2} \\ \vdots \\ \Delta u_{k-m} \end{bmatrix} ; \underset{\leftarrow}{y}_k = \begin{bmatrix} y_k \\ y_{k-1} \\ \vdots \\ y_{k-m} \end{bmatrix} ; \underset{\rightarrow}{y}_{k+1|k} = \begin{bmatrix} y_{k+1} \\ y_{k+2} \\ \vdots \\ y_{k+n} \end{bmatrix} \tag{4}$$

In conventional PFC [6,20], within the predictions we select $\Delta u_{k+i} = 0$, $i > 0$. Uusing this and substituting the n-step ahead prediction from (3) into equality (2) gives:

$$H_n \mathbf{e}_1 \Delta u_k + P_n \Delta \underset{\leftarrow}{u}_k + Q_n \underset{\leftarrow}{y}_k = (1 - \lambda^n)r + \lambda^n y_k \tag{5}$$

where $\mathbf{e}_i$ is the ith standard basic vector and $H_n = \mathbf{e}_n^T H, P_n = \mathbf{e}_n^T P, Q_n = \mathbf{e}_n^T Q$. Hence, the PFC control law is given as:

$$\Delta u_k = \frac{1}{h_n} \left[(1 - \lambda^n)r + \lambda^n y_k - Q_n \underset{\leftarrow}{y}_k - P_n \Delta \underset{\leftarrow}{u}_k \right]; \quad h_n = H_n \mathbf{e}_1 \tag{6}$$

Remark 1. *Figure 1 shows a comparison of simplified flow diagrams, where both PFC and MPC share the same structure, yet have a different optimisation process, where the constraint handling is embedded inside the main block. As for PID, the control input is obtained simply by tuning the gains, while the constraints are handled via a rule base [22].*

Remark 2. *It is noted that PFC performs well for a system with close to a monotonic step response, such as a first-order system and overdamped second-order dynamics, assuming, of course, a sensible choice of the tuning parameters λ and n [11–13]. However, the same tuning procedure may not work for systems with less simple open-loop dynamics, leading to ill-posed decision making and unreliable control.*

PFC/MPC control structure

PID control structure

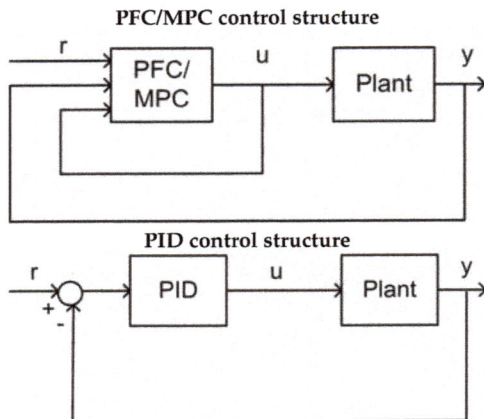

Figure 1. Comparison of flow diagrams for predictive functional control (PFC), model predictive control (MPC), and proportional integral derivative (PID).

2.2. Constraint Handling

The constraint handling approach presented here is adapted from the standard MPC literature [21,23] but with simpler code, and it is more systematic and less conservative than the back-calculation typically used in conventional PFC algorithms [24]. Similarly, it will be more efficient than the ad hoc approaches used with PID. Assume constraints, at every sample, on input rate, input, and states, as follows:

$$\underline{\Delta u} \leq \Delta u_k \leq \overline{\Delta u}; \quad \underline{u} \leq u_k \leq \overline{u}; \quad \underline{y} \leq y_k \leq \overline{y}, \ \forall k \tag{7}$$

The user needs to:

1. Find a suitable horizon m [23] over which to compute the entire set of future output predictions in a single vector: $\underset{\rightarrow k+1|k}{y} = H\mathbf{e_1}\Delta u_k + P\underset{\leftarrow}{\Delta u_k} + Q\underset{\leftarrow}{y_k}$. The horizon for output predictions $\underset{\rightarrow k+1}{y}$, and thus the row dimension of H, should be long enough to capture all core dynamics!

2. Combine the input constraints, rate constraints, and output predictions into a single set of linear inequalities:

$$C\Delta u_k \leq \mathbf{f}_k \tag{8}$$

$$C = \begin{bmatrix} 1 \\ -1 \\ 1 \\ -1 \\ H\mathbf{e_1} \\ -H\mathbf{e_1} \end{bmatrix}; \quad \mathbf{f}_k = \begin{bmatrix} \overline{u} - u_{k-1} \\ -\underline{u} + u_{k-1} \\ \overline{\Delta u} \\ -\underline{\Delta u} \\ L\overline{y} - P\underset{\leftarrow}{\Delta u_k} - Q\underset{\leftarrow}{y_k} \\ -L\underline{y} + P\underset{\leftarrow}{\Delta u_k} + Q\underset{\leftarrow}{y_k} \end{bmatrix}; \quad L = \begin{bmatrix} 1 \\ 1 \\ \vdots \\ 1 \end{bmatrix}$$

where $\mathbf{f}_k$ depends on past data in $\underset{\leftarrow}{\Delta u_k}, \underset{\leftarrow}{y_k}$ and on the limits.

3. The input/output predictions will satisfy constraints if inequalities (8) are satisfied, and thus the PFC algorithm should consider these explicitly.

Next, a single simple loop is utilised to find the Δu_k closest to the unconstrained solution of (6), while satisfying (8).

Algorithm 1. Simple PFC algorithm with systematic constraint handling

---

At each sample:

1. *Define the unconstrained value for Δu_k from (6).*
2. *Define the vector $\mathbf{f}_k$ of (8) (it is noted that C does not change).*
3. *Use a simple loop covering all the rows of C as follows:*

   (a) *Check the ith constraint that is the ith row $\mathbf{e}_i^T C \Delta u_k \leq \mathbf{f}_{k,i}$ of $C \Delta u_k \leq \mathbf{f}_k$.*
   (b) *If $\mathbf{e}_i^T C \Delta u_k > f_{k,i}$, then set $\Delta u_k = (f_{k,i})/[\mathbf{e}_i^T C]$, else leave Δu_k unchanged.*

Remark 3. *For a simple and stable open-loop process and for suitably large m, Algorithm 1 is guaranteed to be recursively feasible and, moreover, to converge to a possible value for Δu_k that is closest to the unconstrained choice [12]. However, this benefit does not apply to systems with more challenging dynamics, such as when the open-loop prediction is divergent.*

3. Input Shaping PFC

This section presents the concept of input shaping to pre-stabilise (or pre-condition) the open-loop predictions of processes with undesirable dynamics. The shaping of input predictions can be done either via explicit pole cancellation or pole shaping, and both methods can be used to formulate a PFC control law which retains a recursively feasible constrained solution. A key issue, however, is whether one deploys FIR or IIR parameterisations of the predictions for the future input increments $\Delta u_{k+i|k}$, $i \geq 0$.

3.1. Pre-Stabilisation of Predictions via Pole Cancellation

For systems with poor open-loop dynamics, the constant input assumption of typical PFC does not provide enough flexibility within the predicted input to both cancel the effect of an undesirable pole and to get nice convergent behaviour [13,17,19]. Thus, it is crucial to first stabilise the prediction before implementing it into a control law [25]. The first step is to factorise the poles in the denominator:

$$\Delta y(z) = \frac{b(z)}{a(z)} \Delta u(z); \quad a(z) = a^-(z)a^+(z) \tag{9}$$

where $a^+(z)$ contains the undesirable poles. Utilising the Toeplitz/Hankel form [21], the future output predictions can be computed as:

$$[C_{a-\Delta}][C_{a+}]\underset{\rightarrow k+1}{y} + H_A \underset{\leftarrow k}{y} = C_b \Delta \underset{\rightarrow k}{u} + H_b \Delta \underset{\leftarrow k}{u} \tag{10}$$

where for general polynomial $f(z) = f_0 + f_1 z^{-1} + \dots + f_n z^{-n}$,

$$C_f = \begin{bmatrix} f_0 & 0 & 0 & \cdots \\ f_1 & f_0 & 0 & \cdots \\ \vdots & \vdots & \vdots & \ddots \\ f_n & f_{n-1} & f_{n-2} & \cdots \end{bmatrix}; H_f = \begin{bmatrix} f_1 & f_2 & \cdots & f_n \\ f_2 & f_2 & \cdots & 0 \\ \vdots & \vdots & \vdots & \ddots \\ 0 & 0 & \cdots & 0 \end{bmatrix}$$

Rearranging prediction (10) in more compact form, we get:

$$\underset{\rightarrow k+1}{y} = [C_{a-\Delta}]^{-1}[C_{a+}]^{-1}[\underbrace{C_b \Delta \underset{\rightarrow k}{u} + H_b \Delta \underset{\leftarrow k}{u} - H_A \underset{\leftarrow k}{y}}_{p}] \tag{11}$$

from which the presence of the undesirable modes are transparent through the factor $[C_{a+}]^{-1}$.

Lemma 1. *Selection of the future input sequence* $\Delta \underset{\to k}{u}$, *at each sample, such that the following equality is satisfied:*

$$[C_b \Delta \underset{\to k}{u} + \mathbf{p}] = C_{a+} \gamma \tag{12}$$

where γ *is a convergent sequence or a polynomial, will ensure that the corresponding output predictions in* (11) *do not contain the undesirable modes in* a^+.

Proof. This is self-evident by substitution of (12) into (11), which gives:

$$\underset{\to k+1}{y} = [C_{a-\Delta}]^{-1}[C_{a+}]^{-1}[C_{a+}]\gamma = [C_{a-\Delta}]^{-1}\gamma \tag{13}$$

so that only the acceptable modes in $a^-(z)$ remain in the predictions, along with any components in the convergent sequence γ. It is noted that this choice automatically includes the initial conditions within $\mathbf{p}$ and thus updates each sample as required. □

Remark 4. *Requirement* (12) *can be solved by a small number of simultaneous equations* [21], *where the minimal-order solution can be represented as:*

$$\Delta \underset{\to k}{u} = P_1 \mathbf{p}; \quad \gamma = P_2 \mathbf{p} \tag{14}$$

for suitable P_1, P_2. *The required dimension of non-zero elements in vector* $\Delta \underset{\to k}{u}$ *corresponds to at least one more than the number of undesirable modes* (n_{a+}), *while the order of* γ *is usually taken as* $n_\gamma = n_p - n_{a+}$, *where* n_p *is the effective dimension of* $\mathbf{p}$ *(which depends upon the column dimensions of* H_b, H_A*).*

To ensure the future manipulated control moves are convergent, while adding some flexibility for modifying the output predictions, the input requirement in (14) can be enhanced to:

$$\Delta \underset{\to k}{u} = P_1 \mathbf{p} + C_{a+} \phi \tag{15}$$

where the vector parameter ϕ denotes the degrees of freedom (d.o.f.) within the predictions.

Theorem 1. *Using the new shaped input* (15) *ensures that the undesirable modes do not appear in the output predictions, irrespective of the choice of* ϕ. *The output predictions are convergent if* ϕ *is finite-dimensional or a convergent infinite dimensional sequence.*

Proof. Substitute input prediction (15) into output prediction (12), and the predictions become:

$$\begin{aligned}
\underset{\to k+1}{y} &= [C_{a-\Delta}]^{-1}[C_{a+}]^{-1}[C_b \Delta \underset{\to k}{u} + \mathbf{p}] \\
&= [C_{a-\Delta}]^{-1}[C_{a+}]^{-1}[C_{a+}\gamma + C_b C_{a+}\phi] \\
&= [C_{a-\Delta}]^{-1}[C_b \phi + \gamma]
\end{aligned} \tag{16}$$

The prediction can be represented with an equivalent z-transform:

$$\underset{\to}{y}(z) = \frac{[1, z^{-1}, z^{-2}, \cdots][\gamma + C_b \phi]}{a^-(z)} = \frac{\gamma(z) + b(z)\phi(z)}{a^-(z)} \tag{17}$$

It is known from Lemma 1 that the contribution from γ gives a convergent prediction, and thus overall convergence is obvious as long as $\phi(z)$ is convergent (or an FIR). □

Remark 5. *Noting the definition of* $\mathbf{p}$ *in* (11), *the n-step ahead output prediction with prediction class* (15) *and* (17) *can be put in a more common form as:*

$$y_{k+n|k} = H_{s,n}\phi + P_{s,n}\Delta \underset{\leftarrow}{u}_k + Q_{s,n}\underset{\leftarrow}{y}_k \tag{18}$$

where $H_{s,n}$, $P_{s,n}$, and $Q_{s,n}$ are suitable matrices, and the additional subscript 's' is used to denote shaping and ϕ is taken to be FIR (equivalently a finite dimensional vector). Note, however, that typically for PFC, ϕ is a scalar. Also, it is easy to show [18] that choosing $\phi = 0$ will automatically give the same input predictions as those deployed at the previous sample, which enables consistency of predictions from one sample to the next.

3.2. Pre-Stabilisation via Pole Shaping

It is known that dead-beat pole cancellation can require aggressive inputs, and the minimal-order solutions to (12) are in effect dead-beat input predictions [16,19]. Although dead-beat solutions are easy to define and thus have some advantages in terms of computation and transparency, in practice, a user may desire a less aggressive shaping that is more implementable in a real system. Alongside this, the popularity of dual mode approaches in the literature [26] is partially because they allow the implied input predictions to converge to the steady state asymptotically, rather than in a small, finite number of steps. Thus, a logical question to ask is whether a smoother solution to (12)—that is, one where the implied solutions for $\gamma(z)$, $\phi(z)$ used in (17) have some poles, say $\alpha(z)$—would work better for PFC.

The mainstream MPC community has focussed on optimal control solutions, but, given that PFC is intended to be simple and low-dimensional, the proposal here is that it is more reasonable to investigate the potential of simple default choices for the asymptotic dynamics $\alpha(z)$ within the input and output predictions. Clearly, this choice can be strongly linked to the target closed-loop behaviour and/or system knowledge.

Proposal 1. *By definition, the integrator has a pole on the unit circle—that is, factor $(1 - z^{-1})$—and, conversely, cancelling the pole as in (12) is equivalent to enforcing a pole on the origin—that is, factor $(1 - 0\,z^{-1})$. Hence, the choice of pole factor $\alpha = (1 - 0.5\,z^{-1})$ represents a simple half-way house trade-off between these two choices.*

Proposal 2. *For a process with significant underdamping, the implied $\alpha(z)$ will have only real poles which are chosen to be close to the real parts of the oscillatory poles. This will reduce the undesirable oscillation in the output predictions, but not change the convergence speed, albeit the input may then be somewhat oscillatory.*

Proposal 3. *For open-loop unstable systems, a simple default solution simply inverts the unstable poles, that is, defining $\alpha(z)$ such that $a^+(z_i) = 0 \implies \alpha(1/z_i) = 0$.*

Lemma 2. *The dynamics $\alpha(z)$ will be present in the predictions if the following Diophantine equation is used to solve the input/output prediction pairing.*

$$b(z)w(z) + \alpha(z)p(z) = a^+(z)\hat{\gamma}(z); \quad p(z) = [1, z^{-1}, ...]\mathbf{p} \tag{19}$$

$$\implies \quad \Delta\underset{\rightarrow}{u}(z) = \frac{w(z)}{\alpha(z)}, \quad \underset{\rightarrow}{y}(z) = \frac{\hat{\gamma}(z)}{a^-(z)\Delta(z)\alpha(z)}$$

Proof. First, note that (19) is equivalent to solving:

$$[C_b C_\alpha^{-1}\mathbf{w} + \mathbf{p}] = C_{a+} C_\alpha^{-1}\hat{\gamma} \tag{20}$$

and, moreover, Equation (20) follows directly from enforcing (12) while assuming $\Delta\underset{\rightarrow k}{u} = C_\alpha^{-1}\mathbf{w}$. Hence, substituting this $\Delta\underset{\rightarrow k}{u}$ into (11) gives:

$$\begin{aligned}
\underset{\rightarrow k+1}{y} &= [C_{a-\Delta}]^{-1}[C_{a+}]^{-1}[C_b\Delta\underset{\rightarrow k}{u} + \mathbf{p}] \\
&= [C_{a-\Delta}]^{-1}[C_{a+}]^{-1}[C_b C_\alpha^{-1}\mathbf{w} + \mathbf{p}] \\
&= [C_{a-\Delta}]^{-1}[C_{a+}]^{-1}[C_{a+} C_\alpha^{-1}\hat{\gamma}] = C_{a-\Delta}^{-1} C_\alpha^{-1}\hat{\gamma}
\end{aligned} \tag{21}$$

It is evident, therefore, that the desired poles are in the predictions for both the input and output. □

Remark 6. *The new requirement* (20) *can be solved similarly to* (12), *where the minimal-order solution for* **w** *and* $\hat{\gamma}$ *are:*

$$\mathbf{w} = \hat{P}_1\mathbf{p}; \quad \hat{\gamma} = \hat{P}_2\mathbf{p} \tag{22}$$

Theorem 2. *A convergent prediction class which embeds both the desired asymptotic poles and some degrees of freedom (d.o.f.) can be defined from:*

$$\mathbf{w} = \hat{P}_1\mathbf{p} + C_{a+}\phi; \quad \Delta\underset{\to k}{u} = [C_\alpha]^{-1}[\hat{P}_1\mathbf{p} + C_{a+}\phi] \tag{23}$$

where convergent IIR or FIR ϕ constitutes the d.o.f.

Proof. This is analogous to Theorem 1 and is based on superposition. The additional component in **w**—that is, $C_{a+}\phi$—necessarily cancels the undesirable poles and gives overall convergent output predictions. So, using (21), then: Added hat on top of gamma in Equation (24).

$$\begin{aligned}
\underset{\to k+1}{y} &= C_{a-\Delta}^{-1}C_{a+}^{-1}[C_b\Delta\underset{\to k}{u} + \mathbf{p}] \\
&= C_{a-\Delta}^{-1}C_\alpha^{-1}\hat{\gamma} + C_{a+}^{-1}C_{a+}^{-1}C_\alpha^{-1}[C_{a+}C_b\phi] \\
&= C_{a-\Delta}^{-1}C_\alpha^{-1}[\hat{\gamma} + C_b\phi]
\end{aligned} \tag{24}$$

□

Remark 7. *By extracting the nth row and noting the definition of* **p** *in* (11), *the n-step ahead prediction from* (24) *can be rearranged in a more general form as:*

$$y_{k+n|k} = h_{n,\alpha}\phi + P_{n,\alpha}\underset{\leftarrow}{\Delta u_k} + Q_{n,\alpha}\underset{\leftarrow}{y_k} \tag{25}$$

for suitable $h_{n,\alpha}$, $P_{n,\alpha}$, $Q_{n,\alpha}$, and it is noted that as is conventional for PFC, ϕ has just a single non-zero parameter in order to retain computational simplicity and to have just a single d.o.f. for satisfying the control law (2).

3.3. Proposed Shaping PFC Control Laws

Since the shaped predictions of (18) and (25) are derived in a general form, two new control laws—Pole Cancellation PFC (PC-PFC) and Pole Shaping PFC (PS-PFC)—can be formulated in a conventional manner after selecting a suitable coincidence horizon n and desired closed-loop pole λ.

[PC-PFC] The d.o.f ϕ is computed by substituting prediction (18) of PC-PFC into equality (2), and thus:

$$\phi = \frac{1}{h_{n,s}}\left[(1 - \lambda^n)r + \lambda^n y_k - Q_{n,s}\underset{\leftarrow}{y_k} - P_{n,s}\underset{\leftarrow}{\Delta u_k}\right] \tag{26}$$

then, the current input increment Δu_k is determined simply by inserting ϕ into the predicted input of (15).

[PS-PFC] The d.o.f ϕ is computed by substituting prediction (25) of PS-PFC into equality (2), and thus:

$$\phi = \frac{1}{h_{n,\alpha}}\left[(1 - \lambda^n)r + \lambda^n y_k - Q_{n,\alpha}\underset{\leftarrow}{y_k} - P_{n,\alpha}\underset{\leftarrow}{\Delta u_k}\right] \tag{27}$$

then, the current input increment Δu_k is determined simply by inserting ϕ into the predicted input of (23).

3.4. Constraint Handling Approaches with Recursive Feasibility

A core advantage of MPC, in general, is that the optimised predictions can be restricted to ones which satisfy constraints; the d.o.f. within the predictions are used to ensure constraint satisfaction. For PC-PFC and PS-PFC, the d.o.f. in the predictions is the variable ϕ. This section gives a brief overview of how the constraint inequalities ensuring (7) depend upon ϕ. We will use PS-PFC and assume that the reader can easily find the equivalent matrices for PC-PFC (for which, in effect, $\alpha = 1$).

Noting the definition of future input increments in (23) and output predictions in (25), the constraints inequalities for (7) can be defined as:

$$
L\underline{\Delta u} \leq C_\alpha^{-1}[\hat{P}_1\mathbf{p} + C_{a+}\phi] \leq L\overline{\Delta u};
$$
$$
L\underline{u} \leq C_{I/\Delta}C_\alpha^{-1}[\hat{P}_1\mathbf{p} + C_{a+}\phi] + Lu_{k-1} \leq L\overline{u}; \tag{28}
$$
$$
L\underline{y} \leq H_\alpha\phi + P_\alpha\underleftarrow{\Delta u_k} + Q_\alpha\underleftarrow{y_k} \leq L\overline{y}.
$$

where $C_{I/\Delta}$ is a lower triangular matrix one ones, and L is a vector of ones with an appropriate dimension (typically a horizon long enough to capture the core dynamics in the predictions). The reader should note that the horizon for the predictions used in (28) will, in general, be much longer than the coincidence horizon used in (27), as one needs to ensure that the implied long-range predictions satisfy constraints. The inequalities can be combined for convenience as follows, although this is not necessary for online coding where efficient alternatives may exist:

$$
C\phi \leq \mathbf{f}_k \tag{29}
$$

$$
C = \begin{bmatrix} C_{I/\Delta}C_\alpha^{-1}C_{a+} \\ -C_{I/\Delta}C_\alpha^{-1}C_{a+} \\ C_\alpha^{-1}C_{a+} \\ -C_\alpha^{-1}C_{a+} \\ H_\alpha \\ -H_\alpha \end{bmatrix} ; \quad \mathbf{f}_k = \begin{bmatrix} L[\overline{u} - u_{k-1}] - C_{I/\Delta}C_\alpha^{-1}\hat{P}_1\mathbf{p} \\ L[-\underline{u} + u_{k-1}] + C_{I/\Delta}C_\alpha^{-1}\hat{P}_1\mathbf{p} \\ L\overline{\Delta u} - C_\alpha^{-1}\hat{P}_1\mathbf{p} \\ -L\underline{\Delta u} + C_\alpha^{-1}\hat{P}_1\mathbf{p} \\ L\overline{y} - P_\alpha\underleftarrow{\Delta u_k} - Q_\alpha\underleftarrow{y_k} \\ -L\underline{y} + P_\alpha\underleftarrow{\Delta u_k} + Q_\alpha\underleftarrow{y_k} \end{bmatrix}
$$

Algorithm 2. [PS-PFC with constraint handling]

At each sample:

1. *Define the unconstrained value for ϕ from (27).*
2. *Update the vector $\mathbf{f}_k$ of (29) (it is noted that C does not change).*
3. *Use a simple loop covering all the rows of C as follows:*

   (a) *Check satisfaction of the ith constraint using: $e_i^T C\phi \leq f_{k,i}$.*
   (b) *If $e_i^T C\phi > f_{k,i}$, then set $\phi = (f_{k,i})/[e_i^T C]$, else leave ϕ unchanged.*

Theorem 3. *In the presence of constraints, Algorithm 2 is recursively feasible where the computed ϕ will not only enforce the input/output predictions to satisfy constraints at the current sample, but will also guarantee that one can make the same statement at the next sample.*

Proof. By definition, the choice of $\phi = 0$ ensures feasibility in the nominal case because the input component $\hat{P}_1\mathbf{p}$ is the unused part of the input prediction from the previous sample, and this is known to satisfy constraints by assumption. One can ensure feasibility at start-up by beginning from a sensible state. $\square$

Readers should note that using the pre-stabilised/shaped predictions is essential for this recursive feasibility result, which is not available for more conventional PFC approaches, for which the implied long-range predictions may be divergent. Thus, this Theorem is an important contribution of this paper.

Remark 8. *Although Algorithm 2 allows recursive feasibility, which is a strong result, ironically, the use of PC-PFC or PS-PFC does not not give any a priori stability and/or performance guarantees in general, which is a well-understood weakness of PFC approaches [7] and a consequence of wanting a very simple and cheap control approach.*

4. Numerical Examples

This section presents several numerical examples of the proposed Pole Shaping PFC (PS-PFC) in handling different types of challenging dynamics, while comparing its performance with the Pole Cancellation PFC (PC-PFC) and conventional PFC (PFC). These examples will highlight:

- the impact of input shaping on the open-loop behaviour;
- the trade-off in the closed-loop performance;
- the efficacy of constraint handling;
- the efficacy on laboratory hardware.

For demonstration purposes, the first three processes with varying dynamics are investigated in a MATLAB simulation environment in Sections 4.1–4.4. The final example in Section 4.5 is on laboratory hardware.

4.1. Description of Case Studies

The case studies presented here are inspired from the real process applications. However, for clarity of presentation, the model parameters are not specific to a given piece of apparatus, but rather are generic to attain suitable dynamics which enable an explicit comparison between the control laws. In this work, it is assumed that there is no plant model mismatch. The robustness properties of these controllers will remain as future work, although it is noted that, as with most predictive controllers, disturbance rejection and offset free tracking is implicit.

4.1.1. Case Study 1: Boiler Level Control

In the process industry, the use of a boiler is frequent, and the level of water needs to be controlled within the manufacturer's specified limits. Exceeding the allowable limits may lead to water overflow, overheating, and/or damage to many components. Conversely, if the level is low, the water wall tubes may overheat and cause tube ruptures, resulting in expensive repairs and other potential hazards. Hence, the prime control objective is to raise the water level at the boiler start-up point while retaining it at a constant steam load. Since the conversion process from water to steam is very slow, a typical model for this process is usually a first-order system with an integrator and stable zero [27]. In a discrete form, one of the poles should reside in a unit circle. The relationship between the output water level (m), $y(z)$, and the input water flow rate (m^3 s^{-1}), $u(z)$, can be represented by a representative model, such as G_1:

$$G_1 = \frac{0.1z^{-1} + 0.4z^{-2}}{(1 - 0.8z^{-1})(1 - z^{-1})} \tag{30}$$

4.1.2. Case Study 2: Depth Control of Unmanned Free-Swimming Submersible (UFSS)

In the marine application, the depth of an unmanned submarine can be controlled by deflecting its elevator surface, whereby the vehicle will rotate about its pitch axis; the associated vertical forces due to the water flow beside the vehicle enable the vehicle to sink or rise. Since a step input deflection may create an oscillatory angle of dive due to the water current, typical dynamics to represent this system often consist of at least one stable pole and two complex poles with stable zeros [22]. Thus a representative third-order underdamped process G_2 can be assumed to represent this pitch control system:

$$G_2 = \frac{0.85z^{-1} - 1.5z^{-2} + 0.85z^{-2}}{(1 - 0.6z^{-1})(1 - 1.6z^{-1} + 0.8z^{-2})} \tag{31}$$

with the output as the pitch angle (rad) and input as the input elevator deflection (m).

4.1.3. Case Study 3: Temperature Control of Fluidised Bed Reactor

A fluidised bed reactor is used to produce a variety of multiphase chemical reactions that are highly exothermic and can be considered as unstable. The reactor bed temperature needs to be controlled by manipulating the coolant flow rate to avoid overheating and other potential hazards. In this case, a drastic change in flow rate will trigger a reaction between the chemicals that releases extra energy and increases the bed temperature. In fact, the change in flow rate needs to follow specific dynamics to avoid this reaction while stabilising the temperature. A reduced control model includes at least one stable pole, typically, and one unstable pole [28] to relate the dynamics between the output temperature ($°C$) and input coolant flow rate (m^3 s^{-1}). Inspired by this system, a representative second-order unstable process G_3 is considered as a good case study:

$$G_3 = \frac{0.2z^{-1} - 0.26z^{-2}}{(1 - 0.9z^{-1})(1 - 1.5z^{-1})} \tag{32}$$

4.2. The Impact of Input Shaping on Predictions and Feasibility

The prime purpose of shaping the future input dynamics is to eliminate the effect of unwanted poles in the future predictions. Nevertheless, it is also undesirable to have an overaggressive input activity, which may not be implementable in a real plant. To analyse this issue, the prediction behaviour of PFC, PC-PFC, and PS-PFC for processes G_1, G_2, and G_3 are plotted in Figure 2. From these results, it can be observed that:

- For an integrating process, such as G_1, the constant input prediction of PFC leads to a divergent output prediction, and thus, output constraints can only be satisfied if the input is selected to be zero! Hence, the PFC plots do not appear in this example, as the constraint handling forces a choice of $u_k = 0, \forall k$.
- For G_1, the default input prediction (Equation (15)) for PC-PFC (blue-dotted line) is of dead-beat form and aggressive, whereas the input prediction (Equation (23)) for PS-PFC moves smoothly to the steady state and is less aggressive. There is no significant difference in the corresponding output predictions.
- For the underdamped system G_2, the output prediction from PFC includes a significant oscillation, which is undesirable and will also cause constraint handling to be conservative. The differences between PC-PFC and PS-PFC predictions are similar to those noted for G_1, that is, PS-PFC has a much smoother and less aggressive input prediction, albeit slow, and output prediction, due to the choice of α. Of course, this difference means that the constraint handling for PS-PFC will be far more preferable and less conservative, in general. Conversely, since PC-PFC cancelled out two of their oscillatory open-loop poles, a sudden spike or aggressive damping is expected in the output response.
- For the unstable process G_3, a conventional PFC cannot be used because the divergent predictions will automatically violate constraints so that no feasible choice for u_k will exist. Once again, it is seen that the predictions for PS-PFC are preferable to those from PC-PFC.

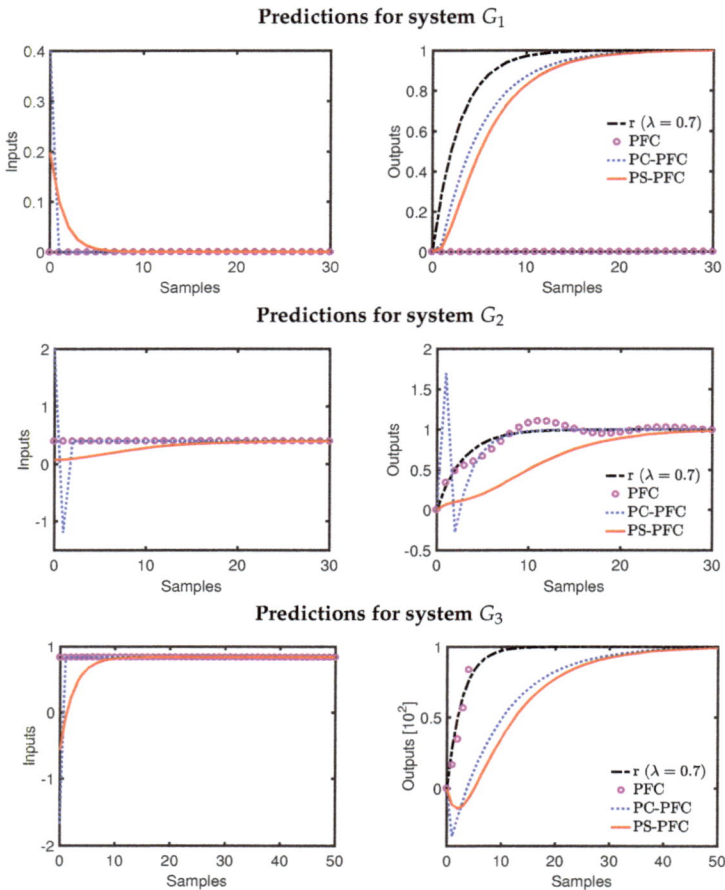

Figure 2. Input and output predictions with PFC, PC-PFC, and PS-PFC for processes G_1, G_2, G_3.

In summary, PS-PFC produces the best prediction behaviour because it ensures convergent predictions which will satisfy constraints with less aggressive input predictions, and thus less conservative constraint handling, than given by PC-PFC/PFC.

4.3. Tuning Efficacy and Closed-Loop Performance: The Unconstrained Case

First we give a brief discussion on PFC tuning for completeness. In general terms, a good practice guidance is to select the coincidence horizon in between 40% and 80% rise of the step input response to the steady-state value [13]: here, G_1 ($4 \leq n \leq 9$), G_2 ($8 \leq n \leq 15$), and G_3 ($11 \leq n \leq 21$). Selecting a smaller horizon will lead to a more aggressive input, while larger horizons reduce the efficacy of λ as a tuning parameter.

In general therefore, the main designer choice is the desired closed-loop pole; for simplicity of illustration, we take the desired pole to be $\lambda = 0.7$. Figure 3 demonstrates the closed-loop performance of PFC, PC-PFC and PS-PFC on the three case study processes with these tunings. It is noted that:

- For all cases, PS-PFC (using a default choice of α) gives the best trade-off between the rate of convergence and the aggressiveness of input activity compared to PFC and PC-PFC. Changes to α could offer a further tuning parameter for varying this trade-off.

- It is notable that PS-PFC gives similar or better output behaviour to PFC/PC-PFC while using a far smoother and less aggressive input trajectory.
- For processes G_2, G_3, the input and output behaviour of PC-PFC is extremely aggressive and would not be implementable in a real application.
- For process G_3, the conventional PFC cannot be stabilised with the given choice of n.

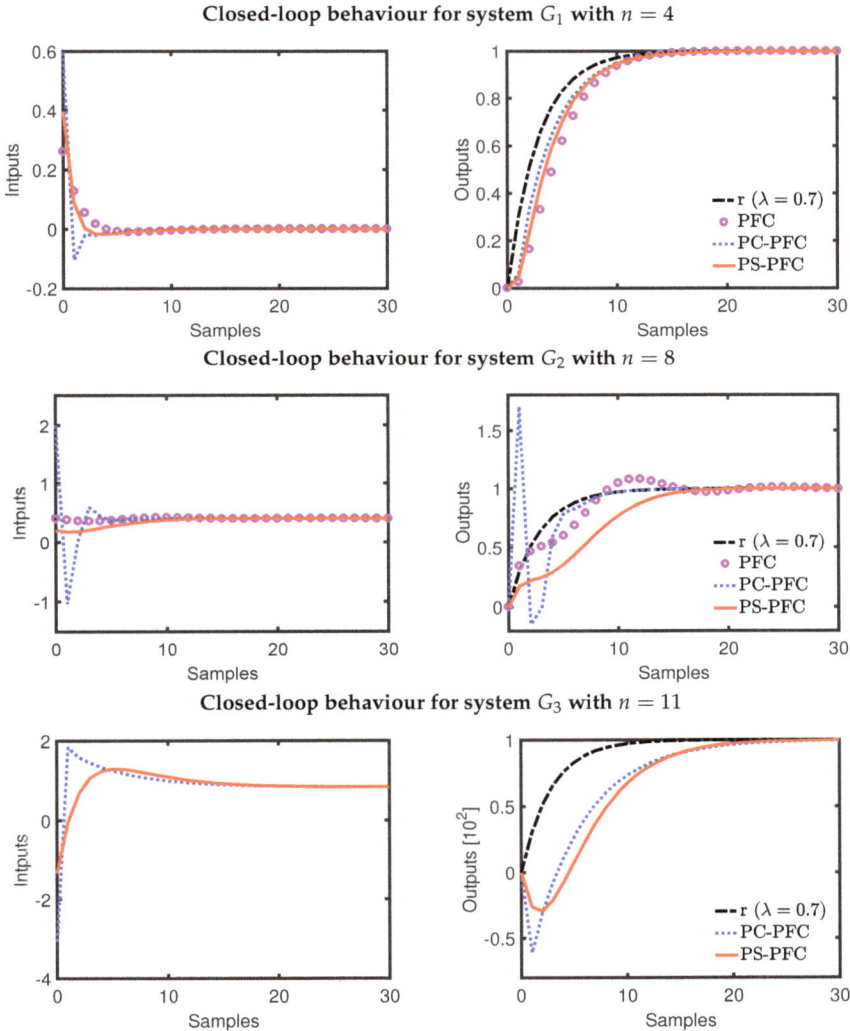

Figure 3. Closed-loop input and output behaviour of PC-PFC and PS-PFC for processes G_1, G_2, G_3.

4.4. Constraint Handling

As noted in Section 4.2, for many dynamics a conventional PFC approach is infeasible or highly conservative because the output predictions inevitably violate constraints beyond a given horizon. Thus, conventional PFC can only be implemented by using short constraint horizons and thus with the loss of any recursive feasibility assurance; if it does work this is luck not good design and thus should be avoided. For the case studies here, conventional PFC could only be used safely with output constraints for G_2, although in that case we would expect some conservatism due to the oscillations in the output predictions.

PC-PFC and PS-PFC pre-stabilise the output predictions and thus can be used safely and with a recursive feasibility assurance. Figure 4 compares the performance of PFC, PC-PFC, and PS-PFC when constraints are added to the process. Several observations can be noted:

- As expected, PS-PFC and PC-PFC satisfy constraints, retain recursive feasibility throughout and converge safely.
- For process G_1, the constrained performance of the controllers are almost the same, but PS-PFC provides a smoother input transition.
- For process G_2 it is clear that handling the under-damping will cause some challenges to any control law, but clearly PS-PFC provides the best responses.
- For process G_3, the inherent dead-beat input predictions deployed by PC-PFC mean the performance is poor and slow to converge whereas PS-PFC performs well.

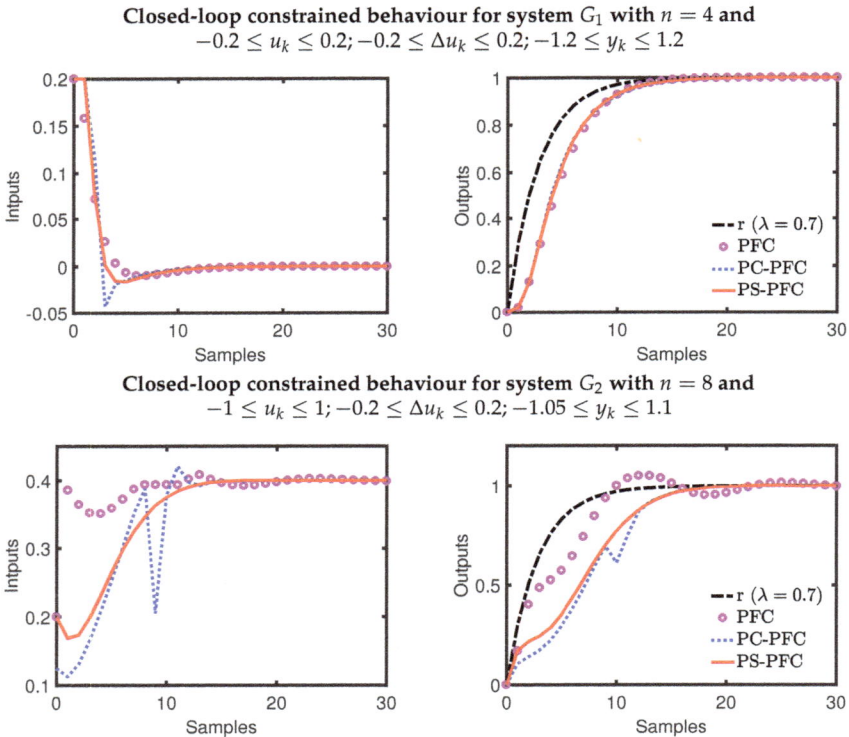

Closed-loop constrained behaviour for system G_1 with $n = 4$ and
$-0.2 \leq u_k \leq 0.2; -0.2 \leq \Delta u_k \leq 0.2; -1.2 \leq y_k \leq 1.2$

Closed-loop constrained behaviour for system G_2 with $n = 8$ and
$-1 \leq u_k \leq 1; -0.2 \leq \Delta u_k \leq 0.2; -1.05 \leq y_k \leq 1.1$

Figure 4. *Cont.*

Closed-loop constrained behaviour for system G_3 with $n = 11$ and
$$-1 \le u_k \le 1; -0.2 \le \Delta u_k \le 0.2; -0.2 \le y_k \le 1.2$$

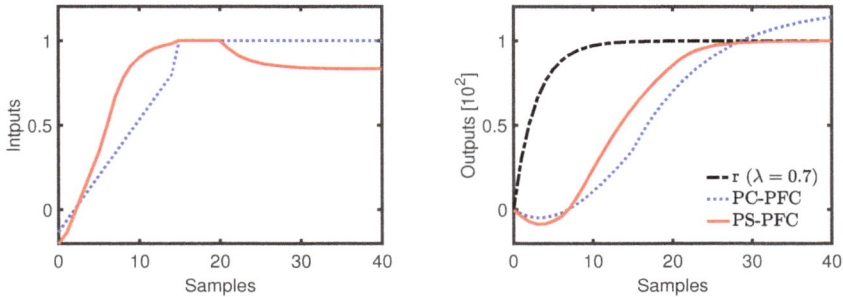

Figure 4. Constrained input and output behaviour of PFC, Pole Cancellation PFC (PC-PFC), and Pole Shaping PFC (PS-PFC) for processes $G1, G2, G3$.

4.5. Application of PS-PFC on Laboratory Hardware

This section demonstrates the implementation of PS-PFC on laboratory hardware, that is, a Quanser SRV02 servo-based unit (Quanser, Markham, ON, Canada) (see Figure 5). This device is powered by a Quanser VoltPAQ-X1 amplifier and operates by National Instrument ELVIS II+ (National Instruments, Austin, TX, USA) multifunctional data acquisition via USB connection. The control objective is to track the servo position $\theta(t)$, measured in radians, by manipulating the supplied voltage $V(t)$. This servo will rotate counter-clockwise with positive supplied voltage and vice versa. A second-order model of (33) with an integrator is used to represent the system dynamics (refer to [29] for a more formal derivation) as:

$$\theta(t) = \frac{1.53}{s(0.254s + 1)} V(s) \tag{33}$$

Figure 5. Quanser SRV02 servo based unit.

To implement the proposed control law, the continuous model (33) is discretised with sampling time $0.02s$ to obtain the discrete model of:

$$G_s = \frac{0.0095z^{-2} + 0.0073z^{-1}}{1 - 1.45z^{-1} + 0.45z^{-2}} \tag{34}$$

The plant is set to have a CLTR of $0.2s$ (which is equivalent to $\lambda = 0.89$). Using a similar procedure to that in the previous section, the coincidence horizon is selected at $n = 4$. Figure 6 demonstrates the unconstrained and constrained performances of PS-PFC in tracking the servo position. In the unconstrained case, the controller manages to track the target position with the desired convergence speed. As for the constrained case, all the implied input limits ($-8 \le u_k \le 8$), rate limits $(-3 \leq \Delta u_k \leq 3)$, and output limits $(-0.8 \leq y_k \leq 0.8)$ are satisfied systematically without any conflict.

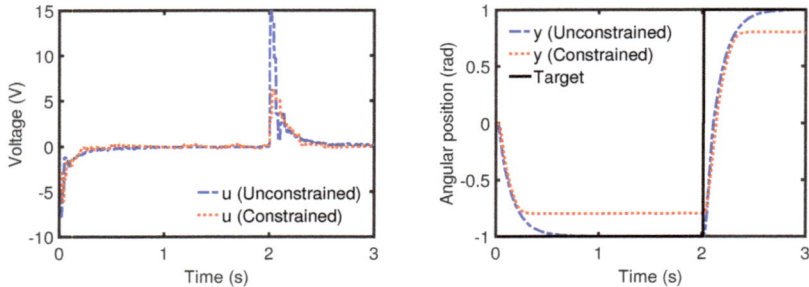

Figure 6. Unconstrained and constrained performance of PS-PFC for process G_s.

5. Conclusions

This paper proposes two potential simple modifications to a conventional PFC algorithm to improve the constraint handling properties for processes with challenging dynamics, such as integrating modes, underdamping, or unstable modes. Both proposals use relatively simple algebra—in effect, the solution of a small number of linear simultaneous equations—to parameterise the future input trajectories which lead to convergent and desirable output behaviours; this is done in terms of a component to deal with the current state (or initial condition) and a free component for control purposes. A core contribution is to show that using the proposed parameterisations allows a simple proof of recursive feasibility so that the constraint handling can be performed more safely and reliably.

A specific novelty of this paper is the proposed PS-PFC algorithm, which gives a pragmatic and simple proposal for deriving input and output prediction pairs which do not require aggressive inputs during transients; the more classical alternative approach of PC-PFC, in general, deploys very aggressive inputs in transients and thus cannot be used in practice. Simulation evidence on a variety of simulation case studies and hardware demonstrates that the proposed PS-PFC algorithm significantly outperforms both a conventional PFC approach and PC-PFC.

Although, as with all predictive control laws, both PC-PFC and PS-PFC are robust to some parameter uncertainty and disturbances, a detailed sensitivity analysis is an important next step. Also, it is of particular interest to compare these approaches with the alternative feedback formulations for PFC [6,15] by way of systematic design, nominal performance, constraint handling, and sensitivity.

Author Contributions: This paper is a collaborative work between both authors. J.A.R. provided initial proposals and accurate communication of the concepts employed in previous MPC and PFC control laws while reviewing the whole project. M.A. developed the code and analysed the concept in various challenging dynamics process within PFC framework.

Funding: This research received no external funding.

Acknowledgments: The first author would like to acknowledge International Islamic University Malaysia and Ministry of Higher Education Malaysia for the scholarship.

Conflicts of Interest: The authors declare no conflict of interest.

References

1. Khadir, M.; Ringwood, J. Stability issues for first order predictive functional controllers: Extension to handle higher order internal models. In Proceedings of the International Conference on Computer Systems and Information Technology, Algiers, Algeria, 19–21 July 2005; pp. 174–179.
2. Richalet, J. Industrial applications of model based predictive control. *Automatica* **1993**, *29*, 1251–1274. [CrossRef]
3. Haber, R.; Rossiter, J.A.; Zabet, K. An alternative for PID control: Predictive functional Control—A tutorial. In Proceedings of the 2016 American Control Conference (ACC), Boston, MA, USA, 6–8 July 2016; pp. 6935–6940.
4. Clarke, D.W.; Mohtadi, C. Properties of generalized predictive control. *Automatica* **1989**, *25*, 859–875. [CrossRef]
5. Haber, R.; Bars, R.; Schmitz, U. *Predictive Control in Process Engineering: From the Basics to the Applications*; Wiley-VCH: Weinheim, Germany, 2011.
6. Richalet, J.; O'Donovan, D. *Predictive Functional Control: Principles and Industrial Applications*; Springer: London, UK, 2009.
7. Rossiter, J.A. A priori stability results for PFC. *Int. J. Control* **2016**, *90*, 305–313. [CrossRef]
8. Zabet, K.; Rossiter, J.A.; Haber, R.; Abdullah, M. Pole-placement predictive functional control for under-damped systems with real numbers algebra. *ISA Trans.* **2017**, *71*, 403–414. [CrossRef] [PubMed]
9. Khadir, M.; Ringwood, J. Extension of first order predictive functional controllers to handle higher order internal models. *Int. J. Appl. Math. Comput. Sci.* **2008**, *18*, 229–239. [CrossRef]
10. Rossiter, J.A.; Haber, R.; Zabet, K. Pole-placement predictive functional control for over-damped systems with real poles. *ISA Trans.* **2016**, *61*, 229–239. [CrossRef] [PubMed]
11. Abdullah, M.; Rossiter, J.A. Utilising Laguerre function in predictive functional control to ensure prediction consistency. In Proceedings of the 2016 UKACC 11th International Conference on Control (CONTROL), Belfast, UK, 31 August–2 September 2016 .
12. Abdullah, M.; Rossiter, J.A.; Haber, R. Development of constrained predictive functional control using Laguerre function based prediction. *IFAC-PapersOnLine* **2017**, *50*, 10705–10710. [CrossRef]
13. Rossiter, J.A.; Haber, R. The effect of coincidence horizon on predictive functional control. *Processes* **2015**, *3*, 25–45. [CrossRef]
14. Richalet, J.; Rault, A.; Testud, J.; Papon, J. Model predictive heuristic control: Applications to industrial processes. *Automatica* **1987**, *14*, 413–428. [CrossRef]
15. Zhang, Z.; Rossiter, J.A.; Xie, L.; Su, H. *Predictive Functional Control for Integral Systems*; PSE: Bellevue, WD, USA, 2018. (In Chinese)
16. Rossiter, J.A. Input shaping for pfc: How and why? *J. Control Decis.* **2015**, *3*, 105–118. [CrossRef]
17. Mosca, E.; Zhang, J. Stable redesign of predictive control. *Automatica* **1992**, *28*, 1229–1233. [CrossRef]
18. Rossiter, J.A. Predictive functional control: More than one way to pre stabilise. In Proceedings of the 15th Triennial World Congress, Barcelona, Spain, 21–26 July 2002; pp. 289–294.
19. Rawlings, J.; Muske, K. The stability of constrained receding horizon control. *IEEE Trans. Autom. Control* **1993**, *38*, 1512–1516. [CrossRef]
20. Richalet, J.; O'Donovan, D. Elementary predictive functional control: A tutorial. In Proceedings of the 2011 International Symposium on Advanced Control of Industrial Processes (ADCONIP), Hangzhou, China, 23–26 May 2011; pp. 306–313.
21. Rossiter, J.A. *A First Course in Predictive Control*, 2nd ed.; CRC Press: London, UK, 2018.
22. Nise, N.S. *Control System Engineering*; John Wiley & Sons, Inc.: New York, NY, USA, 2011.
23. Gilbert, E.; Tan, K. Linear systems with state and control constraints: The theory and application of maximal admissible sets. *IEEE Trans. Autom. Control* **1991**, *36*, 1008–1020. [CrossRef]
24. Fiani, P.; Richalet, J. Handling input and state constraints in predictive functional control. In Proceedings of the 30th IEEE Conference on Decision and Control, Brighton, UK, 11–13 December 1991; pp. 985–990.
25. Rossiter, J.A.; Kouvaritakis, B. Numerical robustness and efficiency of generalised predictive control algorithms with guaranted stability. *IEE Proc. D* **1994**, *141*, 154–162.
26. Scokaert, P.O.; Rawlings, J.B. Constrained linear quadratic regulation. *IEEE Trans. Autom. Control* **1998**, *43*, 1163–1169. [CrossRef]

27. Zhuo, W.; Shichao, W.; Yanyan J. Simulation of control of water level in boiler drum. In Proceedings of the World Automation Congress 2012, Puerto Vallarta, Mexico, 24–28 June 2012.
28. Kendi, T.A.; Doyle, F.J., III. Nonlinear control of a fluidized bed reactor using approximate feedback linearization. *Ind. Eng. Chem. Res.* **1996**, *35*, 746–757. [CrossRef]
29. Apkarian, J.; Lévis, M.; Gurocak, H. *Instructor Workbook: SVR02 Based Unit Experiment for LabVIEW Users;* Quanser Inc.: Markham, ON, Canada, 2012.

Perspective

Natural Products Extraction of the Future—Sustainable Manufacturing Solutions for Societal Needs

Lukas Uhlenbrock [1], Maximilian Sixt [1], Martin Tegtmeier [1,2], Hartwig Schulz [3], Hansjörg Hagels [4], Reinhard Ditz [1] and Jochen Strube [1,*]

[1] Institute for Separation and Process Technology, Clausthal University of Technology, 38678 Clausthal-Zellerfeld, Germany; uhlenbrock@itv.tu-clausthal.de (L.U.); sixt@itv.tu-clausthal.de (M.S.); martin.tegtmeier@schaper-bruemmer.de (M.T.); reinhard.ditz@tu-clausthal.de (R.D.)

[2] Schaper & Brümmer GmbH & Co. KG, 38259 Salzgitter, Germany

[3] Julius-Kühn-Institut, Bundesforschungsinstitut für Kulturpflanzen, 14195 Berlin, Germany; hartwig.schulz@julius-kuehn.de

[4] Boehringer Ingelheim Pharma GmbH & Co. KG, 55216 Ingelheim am Rhein, Germany; hansjoerg.hagels@boehringer-ingelheim.com

* Correspondence: strube@itv.tu-clausthal.de; Tel.: +49-532-372-2355

Received: 24 August 2018; Accepted: 23 September 2018; Published: 1 October 2018

Abstract: The production of plant-based extracts is significantly influenced by traditional techniques and the natural variability of feedstock. For that reason, the discussion of innovative approaches to improve the manufacturing of established products and the development of new products within the regulatory framework is essential to adapt to shifting quality standards. This perspective of members of the DECHEMA/ProcessNet working group on plant-based extracts outlines extraction business models and the regulatory framework regarding the extraction of traditional herbal medicines as complex extracts. Consequently, modern approaches to innovative process design methods like QbD (Quality by Design) and quality control in the form of PAT (Process Analytical Technology) are necessary. Further, the benefit of standardized laboratory equipment combined with physico-chemical predictive process modelling and innovative modular, flexible batch or continuous manufacturing technologies which are fully automated by advanced process control methods are described. A significant reduction of the cost of goods, i.e., by a factor of 4–10, and decreased investments of about 1–5 mil. € show the potential for new products which are in line with market requirements.

Keywords: natural products; phytomedicines; extraction; manufacturing; regulatory

1. Introduction

Patients have accepted traditional herbal medicines for a long time for treatment of mostly minor diseases. An assessment of small-molecule pharmaceuticals which were approved between 1950 and 2010 shows that approximately one third are either natural products or natural product derivatives. Counting the synthetic drugs which were "inspired by nature" increases the count to almost 50%. However, manufacturers of herbal medicinal products suffer from major problems such as increasing market pressure by, e.g., the food supplement sector, increasing regulations, and the costs of production. Due to increasingly strict regulation and approval procedures, innovation is seldom observed, and the methods used in process development are outdated [1].

The history of pharmaceuticals has been plant-based for several thousand years, and may be closely related time- and skill-wise to the origins of wine growing, i.e., over 9000 years in Georgia and not much later in Armenia. Likewise, in Traditional Chinese Medicine (TCM) plant-based pharmaceuticals have been employed for thousands of years to treat patients. More importantly, the use of these natural pharmaceuticals has recently experienced a strong revitalization [2].

Compared to this, only the relatively recent developments, i.e., more or less over the last century, have had an impact on the synthetic chemistry on "pharmaceuticals", which grew substantially during this period. The next wave came from the new field of biopharmaceuticals during the 1980s, which, like the synthetics field, has been heavily attacked over the last couple of decades by the so-called generics/biosimilars. However, it should be kept in mind that the search for new drugs and treatments in both fields has repeatedly benefitted from stimulation by natural products, e.g., Taxol as a cancer drug, which was derived from yew needles [1,3].

Natural products are well established in many branches, like pharmaceutical drugs, nutraceuticals, nutrition additives, cosmetics, flavors, and crop protection agents, as sustainable, biodegradable, green, kosher and halal etc. products which are well accepted by consumers [1,3,4].

Manufacturers of natural extracts have to overcome challenges to keep their products within international markets and/or establish new ones, as regulatory demands and sales prices differ internationally [4–6].

The economy and market shares based on efficacy, one key-aspect which rarely comes up in the discussion of synthetics vs. herbal raw-based product molecules in healthcare application, is the fact that herbal raw-based products, be it in the form of a single compound or a complex mixture (i.e., extract), are by design, or perhaps better, by *origin* biodegradable. This is more than can be said for chemosynthetic compounds and most biosynthetic pharmaceuticals. Especially, when looking at controlled handling, and in particular disposal, of the new generation of synthetics, i.e., the so-called "high potency drugs", production can only be handled under the highest safety conditions; the people involved in production have to work under strict protection. It is obvious that the handling, and especially the disposal, of such therapeutics require significant hazard precautions, which increases costs.

With this in mind, the straightforward and safe handling and disposal of biodegradable herbal-based therapeutics might stimulate consideration of product "fate", like where does it go, where does it stay? Does it contaminate soil, water, or other environmental spheres? These questions should lead to a definition and implementation of clear rules and regulations for handling potentially dangerous products.

Plants undisputedly play an important role as a source for novel molecules and products, ranging from flavors to nutrition to cosmetics and medicines. Plant-based medicines still contribute significantly to human health. Nowadays, 11% of the 252 drugs considered as essential by the WHO (World Health Organisation) are of natural origin [7]. According to Newman and Cragg [1], up to 50% of all approved drugs within the last 30 years came directly or indirectly from natural sources. In the field of cancer treatment, 47% off all small molecule drugs are plant based [8].

In May 2018, BMBF (Bundesministerium für Forschung und Bildung) Germany started a joint international initiative which recognizes that new active substances are urgently needed against infectious diseases; natural sources show enormous potential [9–19]. Over the last decades, the improper application of antibiotics has caused more and more bacteria strains to develop resistance. This has fatal consequences. Every year, about 25,000 patients die due to infections by resistant microbes in Europe alone, as estimated by the WHO [20]. The development of new antibiotics even against resistant microbes is laborious and offers the pharmaceutical industry low commercial incentives. Therefore, over the past 30 years, almost no innovative antibiotic drug has been approved for the market. In the research and development pipelines, hardly any new developments can be seen. Consequently, an impulse must be set for the industry to change this situation regarding research in this field of urgent societal needs. BMBF has committed 500 mil. Euro to support these activities over the next ten years.

The scope of this paper is to provide the reader with an overview of the topic of plant-based product extraction and purification, and to analyze future trends. Therefore, the individual markets are depicted with their individual economic key figures alongside trends in, and resulting demands for, future research.

A detailed overview was published in the position paper of the working group "Phytoextracts—Products and Processes" [4]; thus, only a brief overview is provided in Table 1.

Table 1. Overview of trends, perspectives, market situation, and research demand based on [1] when not stated otherwise.

Category	Agrochemicals	Cosmetics	Aroma, Flavours and Nutrition	Pharma
Market volume	1 Billion USD	200 Billion USD	10 Billion USD	107 Billion USD (forecast 2017) [21]
Market growth	Double digit annual growth rate	Double digit annual growth rate	Double digit annual growth rate Market for nutrition additives decreases Market for aromas grows	Double digit annual growth rate Decline in prescription market Growth in over-the-counter market
Challenges	Market dominated by SMEs as well as global players Small volume/low cost products bulk High cost/low volume niche products	Significant amount of products with natural claims but up to 75% synthetic ingredients; No uniform and binding standards for natural, fair-trade, organic labels	Low cost products (in the order of 1–10 €/kg) Many products with small volumes (100–1000 kg/a)	Most products are OTC Only few blockbusters Restrictive regulatory hamstringed R&D
Medium-term research demands	Efficient total process design for SMEs; Integrate process intensification Methods for SMEs and scale-up of infrastructure to fully integrated manufacturers; Energy efficient and low waste processes for decentralised utilization of natural resources [21,22] *	Efficient ways of finding new natural ingredients [23,24]; * Ensuring sustainability of supply	Apply and adopt more often scCO₂, bio-based solvents *, PHWE Biomass valorization, e.g., carrot, broccoli, artichoke etc. do have 30–80% herbal raw material waste	Speed up of development of herbal raw cell fermentation by omics [25] * Process Analytical Technology for inline-analysis of extraction processes; Parametric defined release at manufacturing of herbal raw extracts; * Homogeneity at production of extracts in large-scale Lyophilisation instead of vacuum-belt drying; Fresh herbal raws instead of dried raw material; HGACP instead of GMP on field incl. extraction media and pomace to be deposited on field again
Long-term research demands	Development of new products	Shift from wild collection to greenhouse or field cultivation in Europe; Energy efficient and low waste processes for decentralised utilization of natural resources [21,22]	Energy efficient and low waste processes for decentralised utilization of natural resources [21,22]	Determination of distribution behaviour of herbal raw ingredients in "single pot model" with herbal raw cell membranes and a gastrointestinal membrane for fast prediction of bioavailable components; Efficacy studies for new herbal raws and products which enable IP protection to cover the costs via patents on the new processes

* alternative cultivation techniques, bio-based solvents, and product development are excluded from this discussion in order to maintain the focus on issues regarding production.

2. Products and Business Models

The status quo and trends in phytomedicines are proposed for discussion from an engineering point of view, taking market and regulatory aspects into account.

Figure 1 describes a typical value chain from cultivation/wild collection over extraction towards tableting/marketing in a typical scale of amount/effort and costs/margins. Finally, about 200 t/year of a natural drug/phyto medicine can be provided to the market.

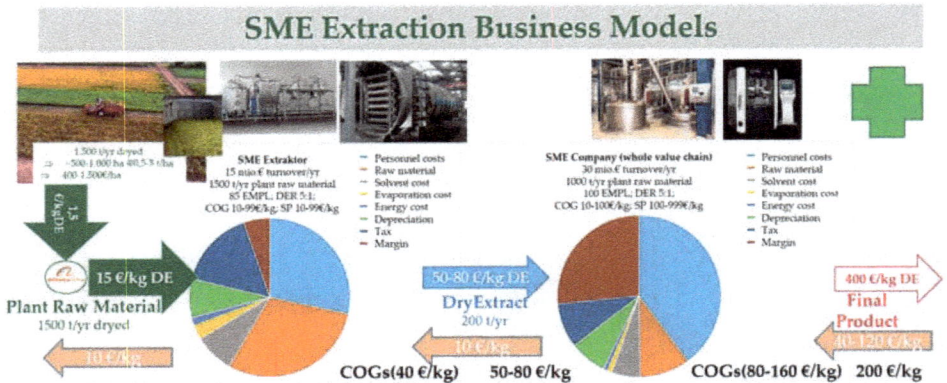

Figure 1. Business models along the value chain: cultivation, extraction, formulation and sales. DE, dry extract; SME, small and medium enterprises; EMPL, employee; DER, drug extract ratio; COG, cost of goods; SP, selling price.

Therefore, about 1500 t/year dried herbal raw material has to be harvested from about 500–1000 ha, assuming a typical harvest of 0.5–5 t/ha of herbal raw species. This results in a yield price of about 400–1500 €/ha typically. The benefit for the farmer is thereby about 1.5 €/kg dry extract for intermediate trade, to be sold further with about 15 €/kg dry extract. Herbal raw trade and cultivation has about 10 €/kg dry extract current revenue from the extraction industry.

The extraction company operates with about 40 €/kg dry extract COGs at the cost distribution as depicted, which is dominated by herbal raw material purchasing costs including solvent and staff. The extraction industry sells to the final formulation and tableting/packaging partners with about 10 €/kg dry extract margin, to which value is added by gaining about 200 t/year dry extract out of about 1500 t/year raw dry herbal raw material. The dry extract has, in the meantime, a value/cost of about 50–80 €/kg dry extract before it is tableted and packed as a phytomedicine, with typical market values of about 400 €/kg dry extract on the consumer market.

Typical extraction companies are SMEs with about 15 mil. € turnover per year, about 85 employees, an average drug extract ratio (DER) of 5:1 at about 1500 t/year dry raw herbal raw material in the order of 10 €/kg dry extract COGs, and a sales price in the same order of magnitude. Typical scales of added value are around 10 €/kg dry extract.

Typical SME phytomedicine companies have a turnover of around 30 mil. € per year, with around 100 employees and 1000 t/year herbal raw material throughput. Average DER of 5:1 results in COGs around 10 €/kg dry extract which have an end sales price in the order of 100 €/kg in relation to the dry extract. The cost distribution of such phytomedicine companies may be seen in the second pie chart; it is dominated by personal, herbal raw material, and energy/solvent. The scale of added value is around 40–120 €/kg, i.e., much higher than the two preceding steps, namely, cultivation and extraction. It is, in any branch, quite usual that the highest value generation occurs close to the end consumer.

3. Regulatory of Herbal Products

In Europe, herbal medicinal products are strictly regulated. Every herbal raw or herbal raw part used for the production of herbal products is described by monographs and is based on assessment reports which are published by the European Medicines Agency's Committee on Herbal Medicinal Products (HMPC) [26,27]. These texts contain information on efficacy, medical indication, toxicological safety, and extract definition. Additional regulations concerning the quality of herbal preparations are established by the European pharmacopeia [28].

The development of new products containing herbal raws, which have not been described yet by HMPC, is rare. Expenses for (pre-) clinical studies are significant, and make product development economically infeasible. Besides, the market application process is time-consuming and fraught with risk, which deters potential investors.

The international approach is not standardized. The US-American Food and Drug Administraion has very strict requirements for the registration of phytomedicines, as described in their legislation "Botanicals" [29]. Generally, it does not recognize herbal products as traditional drugs, but as dietary supplements. According to the FDA, these "supplements are not intended to treat, diagnose, prevent, or cure diseases" [30].

The Chinese Food and Drug Administration (CFDA) licenses herbal products based on 9 different categories [31]:

- Registration Category 1: An active ingredient obtained from herbal raw, animal or mineral materials and its preparations that have not been marketed in China.
- Registration Category 2: A newly-discovered Chinese crude drug and its preparations.
- Registration Category 3: A new substitute for Chinese crude drug.
- Registration Category 4: A new part for medicinal use from currently-used Chinese crude drugs and their preparations.
- Registration Category 5: Active fraction(s) extracted from herbal raw, animal or mineral materials and its preparations that have not been marketed in China.
- Registration Category 6: A combination preparation of TCM or natural medicinal product, which has not been marketed in China.
- Registration Category 7: A preparation with changed administration route of a marketed TCM or natural medicinal product.
- Registration Category 8: A preparation with changed dosage form of a marketed TCM or natural medicinal product.
- Registration Category 9: Generic TCMs or natural medicinal products

This categorization rewards research aimed towards gaining a deeper understanding of the active ingredients and modes of efficacy actions [2,32–40], and not only allows but directly demands the implementation of innovative production techniques as well as actual process development and manufacturing data evaluation methods like QbD- (Quality by Design), combined with PAT- (Process Analytical Technology) approaches for regulatory approval [41,42]. Those technologies are, in the meantime, transferred from biologics to botanicals, successfully applied, and made ready for industrialization at the Sustainable Manufacturing Center for the Chemical-Pharmaceutical Industry at the institute in Clausthal [6,43–47].

With its MIC (Made in China) 2025 strategy [48], China has defined key-technologies which are of crucial national interest. TCM and natural products are of course included. Therefore, the decision has already been made, and substantial resources are already available to gain these aims. Expectations from former comparable actions are that international manufacturers will be reduced to niche markets or need to speed-up their technology and product innovation, which in most cases have been neglected for several decades.

Besides pharmaceuticals, some additional noteworthy aspects of herb-based products are pointed out for other branches:

Herbal based nutritional supplements

- Factor 10–100 larger scale of production compared to pharma,
- regulated environment, but more freedom regarding extraction process,
- semi-purified products

Herb-based herbicides and crop protection

- Factor 100–1000 larger scale of production compared to pharma,
- regulated environment regarding quality, but more freedom regarding extraction process,
- semi-purified products

Cultivation of herbal raw material and resources

- **Pharma**: Regulations specify the origin of herbal raw material; most resources have to be collected from natural habitats, which can problematic because of higher natural variability, environmental impact, risks in supply chain management etc.
- **Nutritional supplements, herbicides, and crop protection:** Cultivated herbal raw material is preferred due to advantages regarding quality, logistics, and secured supply chains.

Quality assurance is based on intensive and expensive laboratory work with a very low degree of automation and data utilization, especially in herbal pharmaceutical production. Innovation regarding inline measurements, data collection and evaluation, and real-time analysis of herbal raw material has the potential to drastically increase the amount of data and ensure stable quality, while maintaining or even reducing current labor costs.

4. Manufacturing Operation of Extracts

Finally, operation of the manufacturing scale has to be discussed for any developed approach. To keep batch variability of natural feedstocks within regulatory demanded requirements is the key challenge.

A fundamental challenge in extracting herbal raws and producing phyto-pharmaceuticals is the natural variety of the feedstock. Therefore, nine different batches differing in harvest year and place were investigated. The overall amount of hyperoside, year, and the country of harvest are summarized in Table 2. Charge I is the reference batch.

Table 2. Overview of Hawthorn lot variety [49].

Lot	Year	Origin	Overall Amount	Deviation Referred to Lot I
A	2017	Southeast Europe	0.87%	146%
B	2016	Macedonia	0.35%	−1%
C	2017	Bulgaria	0.58%	64%
D	2017	Rumania	0.41%	16%
E	2014	Bulgaria	0.60%	71%
F	2017	Albania	0.42%	20%
G	2017	Southeast Europe	0.57%	62%
H	2017	Serbia	0.51%	43%
I	2017	Germany	0.35%	-

The overall amount with respect to the reference batch I varies from −1% to +146%. The percolation runs are depicted below in Figure 2.

To summarize Figure 2, extraction variability due to feed lot variability, such a lot variability, represents the magnitude of statistical variability in that example, i.e., operation points in manufacturing do not influence the natural lot variability with regard to DER and component ingredient content. In contrast, the given approved manufacturing operation point does not influence relevant product quality attributes like DER or component content. Therefore, a regulatory line is discussed in the following section.

Figure 3 depicts a flowchart describing the process from extraction to solvent/auxiliary evaporation/recycling, granulation, and formulation/tableting. Influencing parameters are listed for each unit operation. Easily derived from that overview, critical product quality attributes are only gained at the last tableting step. The effects of the preceding process steps do not influence the criteria for tableting ability, stability, and overall content.

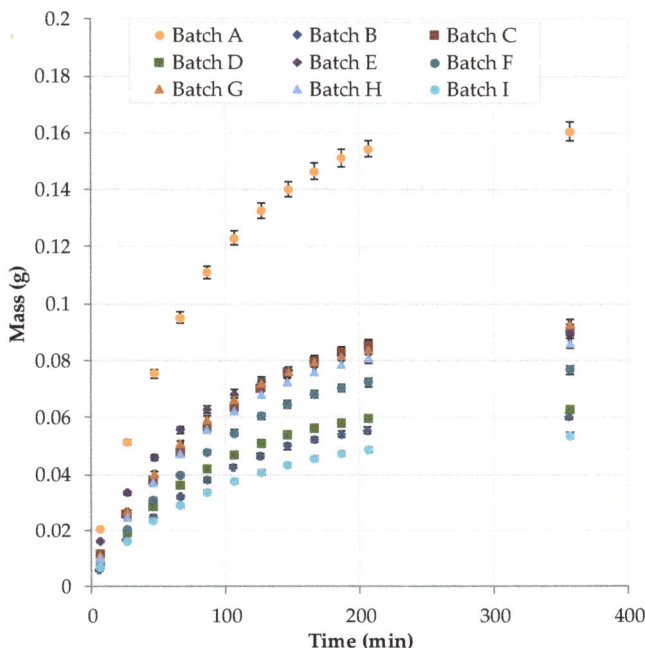

Figure 2. Percolation of the various hawthorn lots [49].

A factor of 2 in feed material content with factor 2 in DER range results in factor 4 in product content; see Figure 4. This may over or under cut the therapeutic ranges, i.e., need to define a therapeutic index, which is, as a standard of modern efficacy-based medicines, quite narrow.

At the typical DER-ranges (4–7.5:1) described in a Pharmacopeia or package insert, and typical content deviations of 0.1–0.5% due to nature, harvest region/time/season/weather, storage/transportation/pre-treatment etc., the range of deviation from active component content is significantly high. Moreover, during temperature-elevated operations like PHWE or SFE or strong active solvents types, components tend to decompose with increasing residence time.

A deviation of factor 4 in content is simply reached by factor 2 in herbal raw material and factor 2 in DER range definition. Due to the fact that this deviation in content of drug charges is statistically obvious, i.e., that efficacy related to at least any therapeutic index, there are significant patient groups which did not get any drugs with enough content for any efficacy. Proof of the efficacy of potential drugs by clinical trials is due to the fact that competition with existing therapies causes too many adverse effects occur, because there is too much content of other/different components inside.

Figure 5 exemplifies the influence of manufacturing parameter during extraction and formulation, tableting on critical product quality attributes.

The effects of manufacturing parameters for critical product quality attributes ends at formulation and tableting. Extraction operation does not effect those quality attributes if products are defined as *extracts*. Only if standardized or quantified extracts are approved does extraction operation has an impact on CQAs, and it is of interest to be controlled in order to gain CQAs robustly at typically changing natural feedstocks.

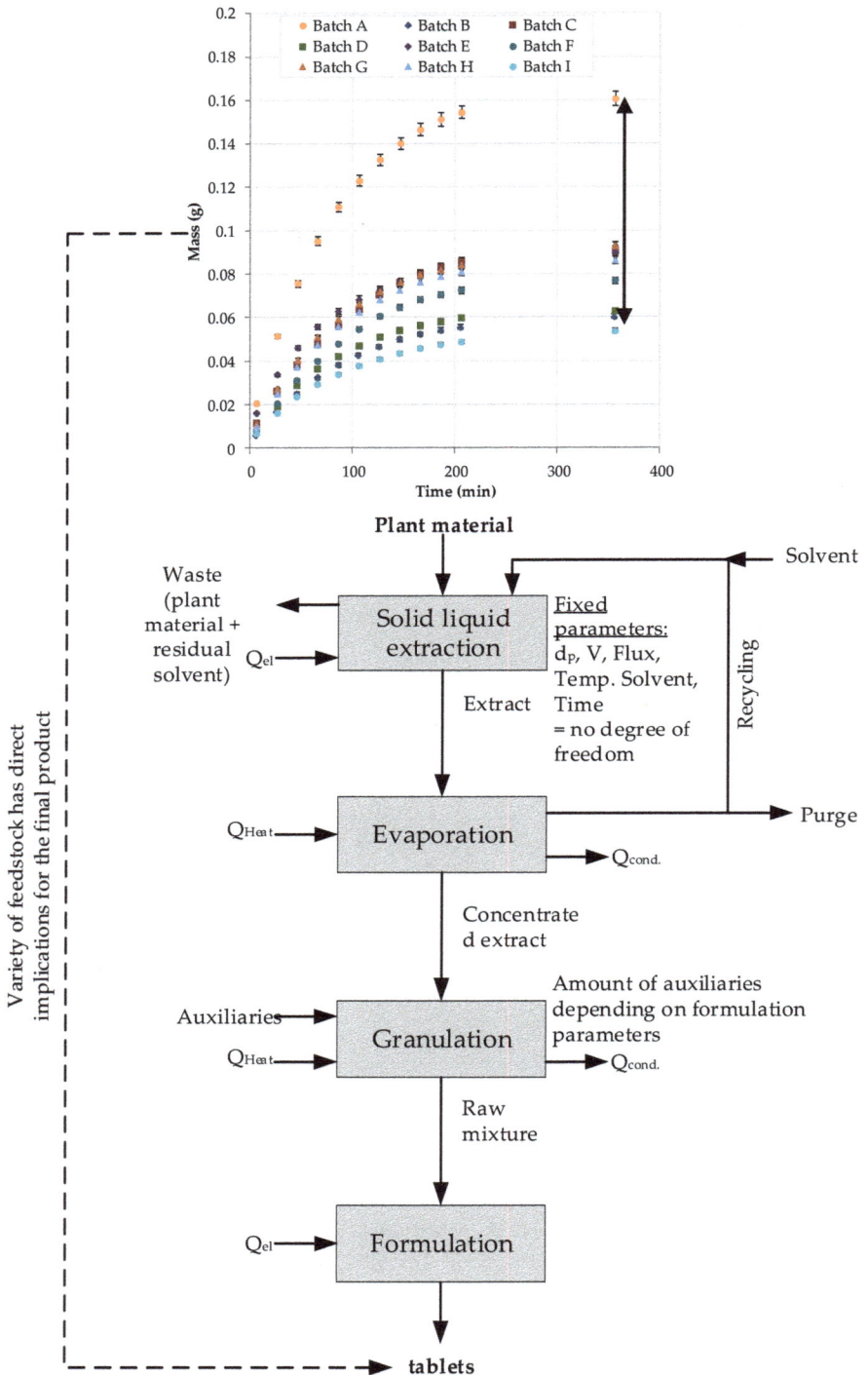

Figure 3. Basic scheme of phytoextraction processes [50].

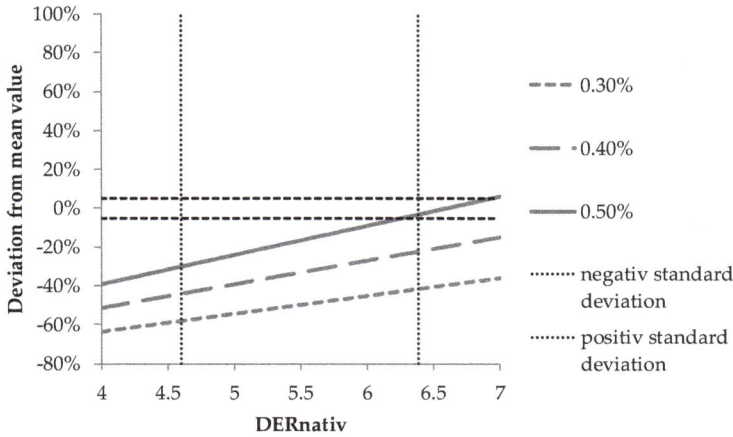

Figure 4. DER deviation derived by herbal raw content nature and definition range [50].

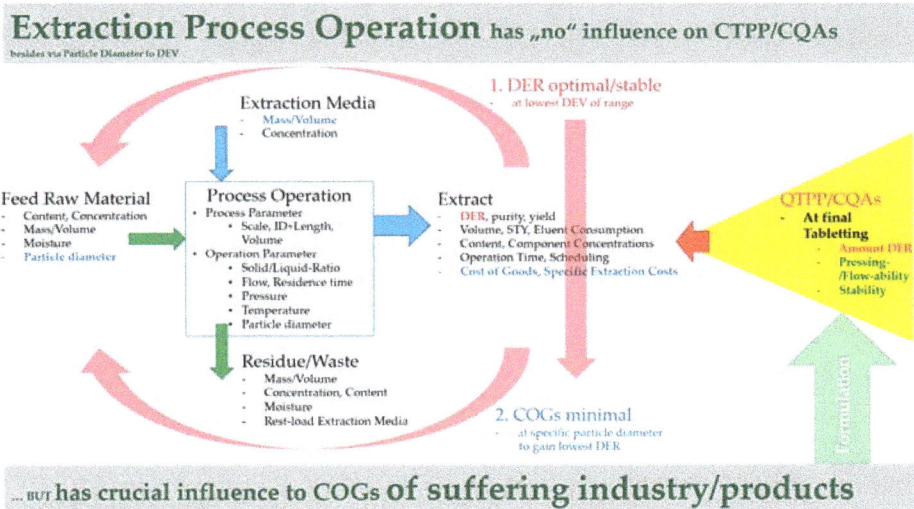

Figure 5. Analysis of root cause for Qbd-approach derived critical product quality attributes effects during manufacturing.

In order to improve other extracts towards quantified extracts, the following work steps are necessary:

- Analytical fingerprint, characterization, lead substances definition
- Efficacy study of lead substances range
- QbD approach to determine operation parameter and
- Submission of Design Space, i.e., new approval
- Economic optimal operation at maximal therapeutic value product

Fingerprinting is applied for any manufacturer with approval documentation in order to document batch specific recovery. Why not officially use this existing data for product and process improvement?

To empower quantified extracts towards standardized extracts, the following work packages are necessary:

- Analytical quantification (not characterization)
- Lead efficacy substance value (not range) i.e., efficacy studies
- Design Space by QbD approach much narrow i.e., robust process i.e., reliable product
- New approval

A discussion of the efforts and benefits of working out the final possible work flow step, approving a purified API:

- The steps described above, with efficacy based on single substances or well characterized substance groups
- Purification process development
- New approval

This results in high efficacy and high product quality (to be analytically distinctly proven) with the highest margins, and with less marketing necessary.

Consequently, it could be derived that the efforts are almost the same magnitude, which requires additional studies aiming at the potentially highest value product.

Financial support by funding organizations could be gained, if wished. The obstacles could be only missing or a lack of technology. This could be overcome by organizational means and/or co-operation, as those technologies are available, as described before.

5. Process Design Proposal for Efficient Manufacturing

Robust process operations generating reliable product quality given the natural variability of herbal raw feedstock is developed at the early stage of the process design and development. Therefore, efficient tools should be available for engineering tasks based on laboratory scale experiments. This tool box has already been proposed about a decade ago, and has been applied to many examples. It has also been developed further to extend technical readiness for daily project work [5,6,43–45,49–59]. In the following section, only a short summary is given to outline the general approach.

5.1. Modelling of the Extraction Process

The aforementioned extraction model assumes an average particle diameter of herbal raw feed material. Therefore, for a particle size distribution (PSD), uniform diffusion paths result. This is a simplifying assumption, which has to be proven at any industrial application first, as the particle size distribution is an easily accessible dimension. Due to this, the model is normally extended to a realistic particle size distribution.

This particle size distribution is implemented within the DPF-model by a summation of the different mass transfer amounts of each individual particle size class of a Q_3-distribution. Parameter γ_i is the mass amount of each class in relation to the total feed herbal raw amount.

$$\frac{\partial c_L(z,t)}{\partial t} = D_{ax} \cdot \frac{\partial^2 c_L(z,t)}{\partial z^2} - \frac{u_z}{\varepsilon} \cdot \frac{\partial c_L(z,t)}{\partial z} - \sum_{i=1}^{n} \gamma_i \frac{1-\varepsilon}{\varepsilon} \cdot k_{fi} \cdot a_{Pi} \cdot [c_L(z,t) - c_{Pi}(r_i = R, z, t)] \quad (1)$$

Fluid dynamic behavior is unaffected by this. The relevant characteristic numbers Péclet and Reynolds, as well as axial dispersion, are determined with the mean particle size diameter $d_{P,mean}$.

Regarding mass transport, the specific surface of the particles, as well as Sherwood and Schmidt numbers, are determined for each class with the following Equations (2)–(5).

$$a_{P,i} = \frac{6}{d_{P,i}} \quad (2)$$

$$Sc = \frac{\eta}{\rho_L \cdot D_{12}} \tag{3}$$

$$Sh_i = \frac{k_{fi} \cdot d_{Pi}}{D_{12}} \tag{4}$$

$$Sh_i = 2 + 1.1 \cdot Sc^{0.33} \cdot Re^{0.6} \tag{5}$$

Figure 6 exemplifies the experimental efforts of model parameter determination at the laboratory scale with increasing modelling depth for the rapid development over the last 10 years or so, starting with shrinking-core and pore diffusion effects taken into account [52,54], diffusion interference of different molecules by Maxwell-Stefan [55,60], shrinking and swelling mass transport kinetics of non rigid herbal raw matrices during extraction up to particle size distributions, and degradation kinetics caused by temperatures [46] up to a recent development, the spectroscopy-assisted model parameter determination [44,61].

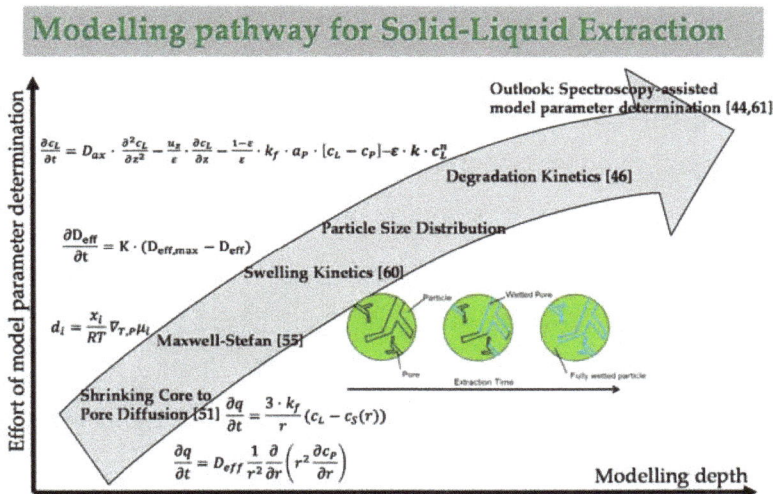

Figure 6. Effort of experimental model parameter determination at increasing modelling depth.

Figure 6 shows the stepwise approach of model equation assembly based on stepwise experimental model parameter determination in laboratory scale in order to develop physico-chemical (rigorous) models, which are a priori predictive because of the effects of fluid-dynamics, mass transfer, phase equilibrium, and energy balances are separated from each other. This general approach was first proposed by Altenhöner for chromatography, and has successfully been transferred to solid-liquid extraction modelling [51].

Figure 7 clearly points out the two boundary cases of mass transfer, related to maceration and percolation process design. Either the mass transfer kinetics are slow such that enough residence time has to be provided in order to gain sufficient total extraction yield, or the capacity of the chosen extraction solvent limits the extraction yield because not enough solvent is provided by the operation parameter window chosen by process design, as shown in Figure 8.

Highly optimized extraction procedures with regards to solvent consumption, recycling effort, yield, robustness, and productivity are especially important for products in the low-price and OTC market segment. High-price products like APIs can normally compensate for unfavorable process designs due to their high margins. This is not desirable at all, due to the waste of natural resources on the one side, and on the other side, best in class processes will complicate market entry for competing companies, so additional money might be spent to maintain the advance in manufacturing and knowledge instead of marketing effort.

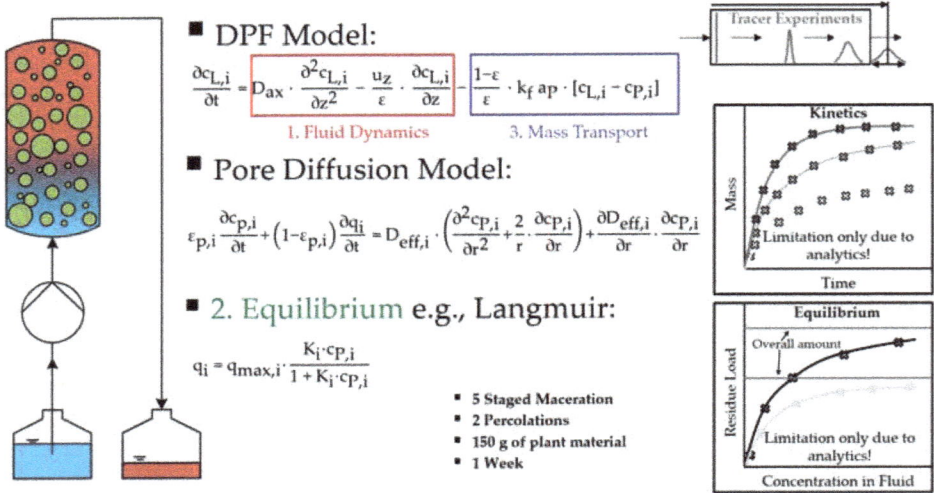

Figure 7. Equation assembly for experimental model parameter determination.

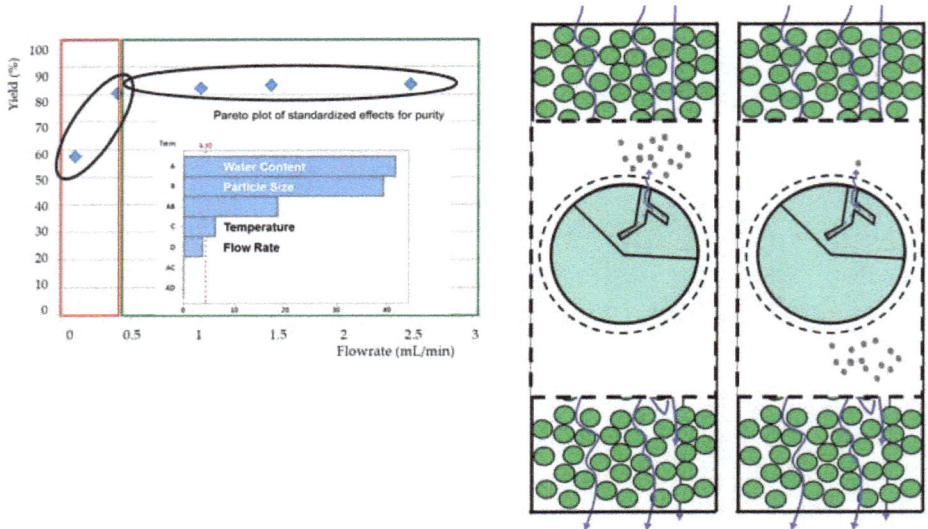

Figure 8. Residence time or solvent capacity limitation—maceration or percolation.

Tables 3 and 4 summarize as a review typical herbal raw material, which have been in process development and design, as well as process modelling and piloting studies, over the last years. From evaluation of this database, some general guidelines for process design and operation of resource-efficient extraction manufacturing can be derived:

1. If a targeted component is within the inner plant particle (e.g., yew, whitethorn, bearberry), then it is recommended that small particles be used to minimize the flow rate required to achieve high

extraction yields. In this way, diffusion limitations can be minimized, and a highly concentrated extract can be obtained.

2. Particle size is not significant if the target component is located on the outer surface (e.g., mugwort, salvia). As a consequence, high flow rates can be used to utilize the fast extraction kinetics. Overestimation of extraction kinetics can result in diluted extracts and a waste of solvent.

3. The extraction of oil requires breakage of oils seams of the particle. The phase equilibrium is practically immeasurable, because only solution mechanisms with extremely fast kinetics occur (e.g., fennel, caraway).

5.2. Resource Efficiency Optimization

The obvious way to reduce costs for solid-liquid extraction is to limit the solvent consumption by minimizing the total extraction time at constant yield. One possibility to achieve this is to shorten the diffusion pathway and to enlarge the specific surface area of the herbal raw material by grinding it to small particles. Other methods, e.g., ultrasound, microwave, or high pressure extraction are possible and have been applied at the lab-scale, but are currently not utilized in industrial production due to economic or technological challenges associated with scale-up [5,6,56].

The pore diffusion model of Section 3 is solved for each particle size class in parallel, and therefore, an individual diffusion path length is calculated for each class. The particle size distribution of the factions of yew needles which are utilized for model validation experiments are visualized in Figure 9. The coarse material shows a monomodal distribution around a median particle diameter of 1010 μm, while the fine material is distributed bimodally.

Figure 10 compares two different extraction progresses between measurements (data points) and simulations (continuous lines). Gray lines are the simulations with the particle collective characterized only by a mean particle diameter. If the fine material is simulated, the prediction with the simplified model assumptions is more accurate. The bimodal character of the distribution results, in the corresponding simulation (black line), in an area where the material is very rapidly leached out. In contrast, large particles delay the extraction and make the extraction progress much slower, as the simulations only with a x_{50}-value show. Simulating the extraction progress of the coarse material, all parameters besides particle size/particle size distribution and the mass of the herbal raw feed material are kept constant. In the described application, the simplified model fails totally to predict the extraction progress, whereas the detailed model, which considers particle size distribution, results in adequate prediction accuracy.

The sketched example exemplifies an extreme operation point. In earlier studies, the influence of varying particle sizes has been less drastic, and predictive simulations were possible [53,55,60]. The extracted particle collectives were distributed almost monomodally, and simulation using the x_{50}-values was an applicable simplification. Moreover, in the majority of those applications, the component of interest for extraction is located on the outer surface of the leaves, e.g., mugworth. Due to this, particle size reduces sensitivity, because diffusion paths within the particle play minor role. If complex particle size distributions are on hand and the active components are homogeneously distributed within the particles, then the particle size distribution of feed materials should be taken into account in order to improve the accuracy of the simulation's predictions. Notably, the PSD is quite simple to determine experimentally.

The extraction simulation shown in Figure 11 demonstrates this. Originating from a mean particle diameter of 1.5 mm, a yield of 80% is reached after 180 min of extraction. The same yield is reached after only 45 min if the particles are ground to 0.1 mm. Thus, the extraction time and the solvent consumption are both reduced by a factor of 4.

Table 3. Database of the physical properties of various herbal raw material for process design.

Category	*Taxus baccata* L.	*Crataegus monogyna* JACQ.	*Foeniculum vulgare* L. Mill.	*Carum carvi* L.	*Artemisia annua* L.	*Arctostaphylos uvaursi* (L.) SPRING.	*Azadirachta indica* A. Juss.
Use	Pharma (cancer treatment)	Pharma (extract for cardiac insufficiency)	Aroma/fragrance	Aroma/fragrance	Pharma (Malaria treatment)	Pharma (extract for bladder infection)	Agro (pest control)
Target component	10-Deacetylbaccatin III (0.1–0.4% w)	Hyperosid (0.3–0.7% w)	Anethole (5.3%), Fenchone (2.9%) essential oil (~8% w)	Carvone, Limonene essential oil (~2% w)	Artemisinin (~0.4% w)	Arbutin (~15% w)	Azadirachtin
Molecular structure and weight							
	544.59 Da	464.38 Da	148.2/152.23 Da	150.22 Da	282.33 Da	272.25 Da	720.71 Da
Side component	Unknown	Unknown	Estragole (0.2%)		Unknown	Hydroquinone	Unknown
Molecular structure and weight							
			148.2 Da			110.11 Da	
Location	Needle Diffusion limitation	Leaf Diffusion limitation	Fruit Oil channels	Fruit Oil channels	Trichoma cells [62]	Leaf	
Solvent	Acetone/Water (80/20 v/v) [43] PHWE (120 °C, moderate decomposition) [46]	Ethanol/Water (70/30 v/v) PHWE (140 °C, no decomposition) [50]	Ethanol [60]		Acetone PHWE (80 °C, fast decomposition) [59,60]	Water (25 °C, fast degradation during maceration) PHWE (140 °C, no decomposition) [50]	Water (25 °C pH 4)

Table 3. Cont.

Category	Taxus baccata L.	Crataegus monogyna JACQ.	Foeniculum vulgare L. Mill.	Carum carvi L.	Artemisia annua L.	Arctostaphylos uvaursi (L.) SPRENG.	Azadirachta indica A. Juss.
Modelling	Pore diffusion	Pore diffusion	Broken Cells	Broken Cells	Film diffusion	Pore diffusion	Pore diffusion
Equilibrium							
Optimization	small dp [60]	small dp [50]	Hydro distillation preferred	Hydro distillation preferred	high flow rate [6]	[50]	
Purification	Benchmark, lab-scale [43]	None	Conceptional	None	Benchmark, Lab-scale, Pilot-scale [50]	None	Lab-scale, Pilot-scale
Basic research	FTIR process control, Raman-mapping [61]	Lot variety [50]	Inline spectroscopy, APC, Raman-Mapping [63]	CLSM, FTIR	Process integration crystallization [50]	Lot variety, decomposition	

Table 4. Database of the physical properties of various herbal raw material for process design (continued).

Category	Vanilla planifolia Jacks. Ex Andrews	Piper nigrum L.	Camellia sinensis (L.) KUNTZE	Salvia officinalis L.	Beta vulgaris L.	Zea mays L.	Larix decidua Mill.
Use	Aroma/Food	Aroma/Food	Aroma/Food	Food (preserving agent)	Food	Food	Agro (pest control)
Target component	Vanillin (3–7% w)	Piperine (~6.5%)	Caffeine (3–6%)	Carnosol (0.1% w)	Sucrose (14–20% w)	Tricin (55 ppm)	Larixol Larixylacetat
Molecular structure and weight	152.15 Da	285.34 Da	194.19 Da	330.42 Da	343.3 Da		306.49 Da

Table 4. *Cont.*

Category	*Vanilla planifolia* Jacks. Ex Andrews	*Piper nigrum* L.	*Camellia sinensis* (L.) KUNTZE	*Salvia officinalis* L.	*Beta vulgaris* L.	*Zea mays* L.	*Larix decidua* Mill.
Side component			Polyphenoles	Carnosoic acid (1.7%)	Ions (Mg, Na, K) Proteins		Tannines
Molecular structure and weight			>1000 Da	332.42 Da			
Location		Fruit	Leaf	Trichoma cells, Film diffusion limitation	Tuber		Bark
Solvent	Ethanol [51]	Ethyl acetate [51]	Water, Ethanol, US Extraction [36]	Acetone [60]	Water [54]		Ethanol/Isopropanol
Modelling	Pore diffusion	Pore diffusion	Pore diffusion	Film diffusion	Pore diffusion	Pore diffusion	Pore diffusion
Equilibrium							
Optimization	[51]		small dp	high flow rate [60]	small dp [9]		
Purification	None	None	None	Benchmark, Lab-scale [60]	None	Conceptional	None
Basic research	Raman-mapping, crystallization [64]	CLSM [51]					

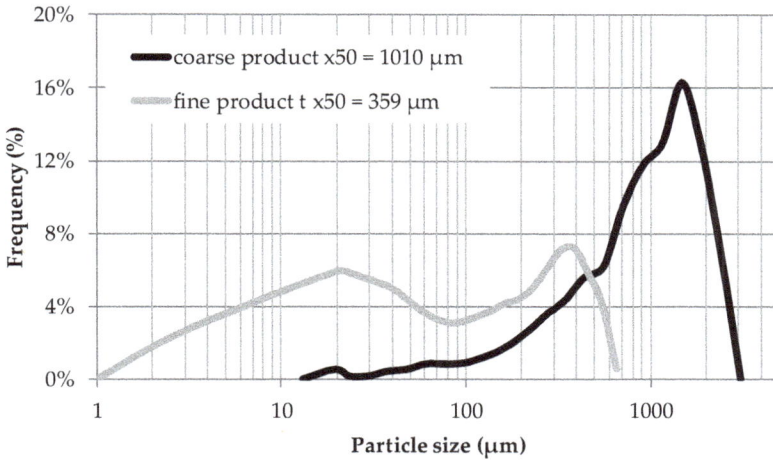

Figure 9. Particle size distributions of different feed herbal raw material batches.

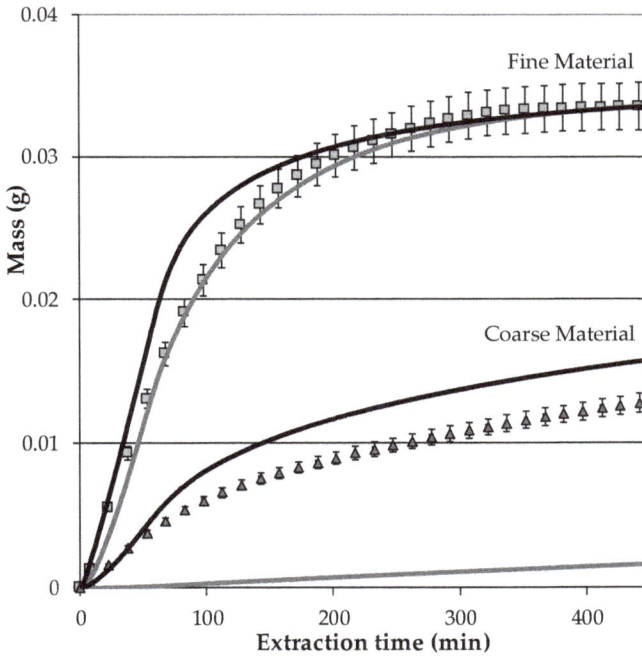

Figure 10. Extraction progress, simulation (continuous lines), and experiments (points) for two different particle size distributions, (black line: simulation with PSD and grey line: simulation with mean particle diameter).

Of course, the optimal process is almost never the sum of the single unit operation optima. Therefore, total process integration and optimization of the complete process chain yields the main profit, as proven in various studies, (e.g., [43,65]). These studies lead to the consequence that particle size, as an essential parameter for extraction, should always be integrated into manufacturing operations, as proposed in the following section.

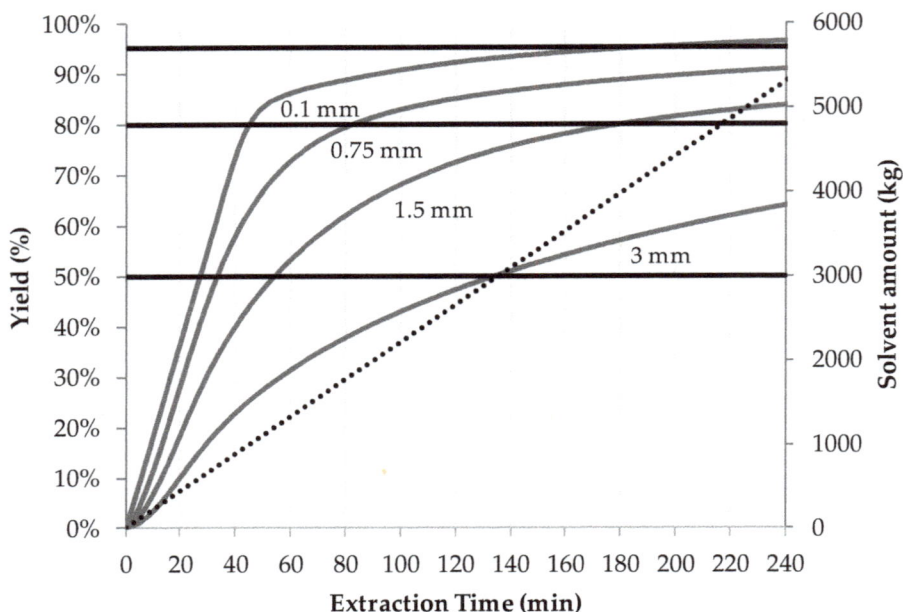

Figure 11. Yield and Solvent Consumption for hawthorn at different mean particle diameters [35].

6. Integrated Continuous Pre-Treatment and Extraction (iCPE)

One key parameter for fast and exhaustive extraction is a small particle diameter due to the higher specific surface area of the ground herbal raw material. The tremendous influence is shown by simulation studies in Figure 12. Different extraction curves considering mean particle diameters from 0.05 mm to 2 mm pare simulated. After 600 min of extraction, only about 20% yield compared to the overall amount of the considered target component is reached at 2 mm. The productivity of the extraction increases significantly with smaller particles; thus, a 95% yield is reached within 110 min using particles with a mean diameter of 0.05 mm. At the same time, the solvent quantity is reduced by 3%.

By applying small particle diameters in extraction, the productivity can be increased drastically, while extraction time and solvent consumption decrease. For that, a continuous operation mode is needed to avoid a bottleneck shift from a slow extraction towards an inefficient handling in between different batches.

6.1. Integrated Continuous Pretreatment and Extraction (iCPE) Process

According to the flow scheme, Figure 13, the herbal raw material is first chopped with a crusher, and then further ground with a ball mill. The particles are constantly blown out of the mill by a fan and transported to a zigzag separator. In the separator itself, a separation takes place, and the large particles are sent back to the mill, purging out a certain quantity. The small particles pass an in-line particle imaging system that controls the mill and the fan in order to get the desired mean particle diameter and distribution. The particles are blown into a percolation column equipped with a pivoted frit at each side. For column filling, the lower frit is closed; therefore, the particles are segregated into the column while the air passes through the frit (Step1). After loading is complete, the column is closed and extraction takes place (Step 2). In the meantime, another column, being in stand-by mode during Steps 1 and 2, is swapped into the filling-position, thus ensuring a continuous operation mode (Step 3).

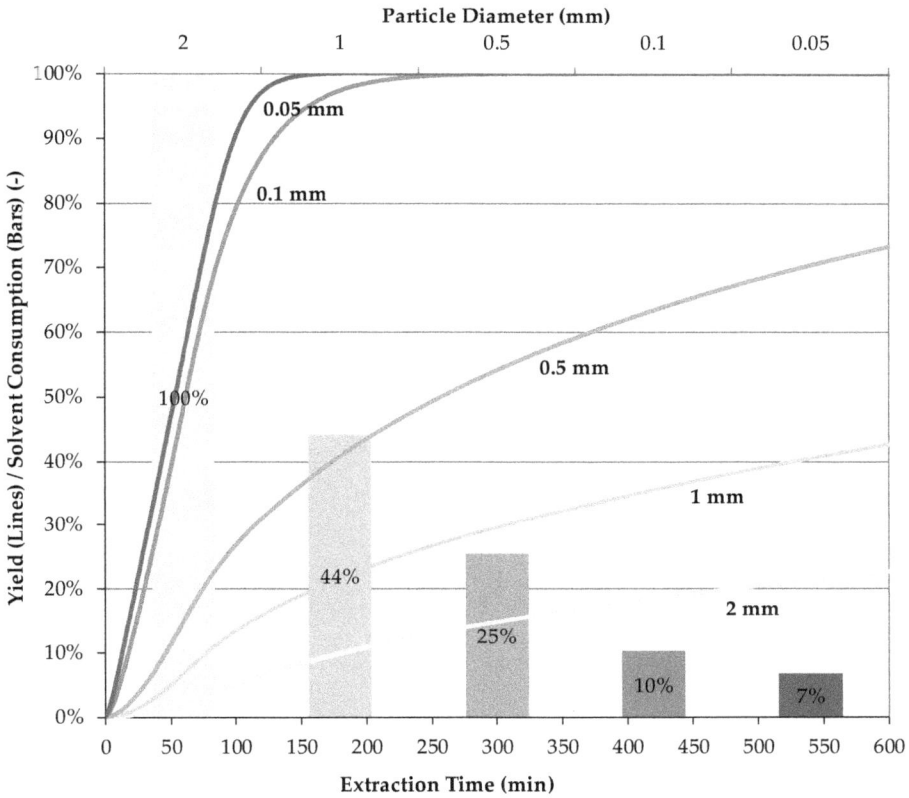

Figure 12. Simulated extraction curves and relative solvent consumptions for different mean particle diameters.

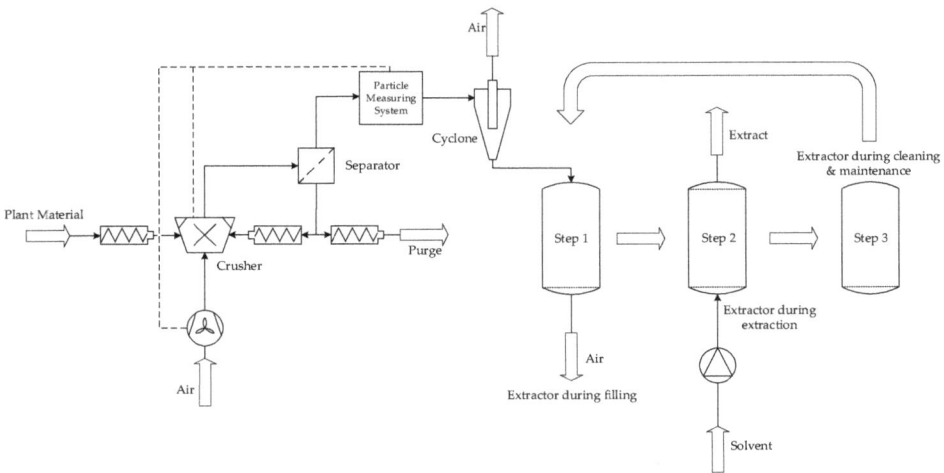

Figure 13. Process proposal for the Integrated Continuous Pretreatment and Extraction (iCPE).

6.2. Cost Calculation and Results of the iCPE Process

A comparison of four different processes is presented in Figure 14. The benchmark is based on a case study for the extraction of 10-deacetylbaccatin III from yew (*Taxus baccata* L.). The extraction takes place in one single 20 m^3 percolation column for 2 h, without milling. Afterwards, a 2 h period of refilling is required. In order to recycle the solvent (a mixture of acetone and water), it is partly evaporated with an agitated film evaporator. These two steps are considered as the standard apparatus for this special case [43].

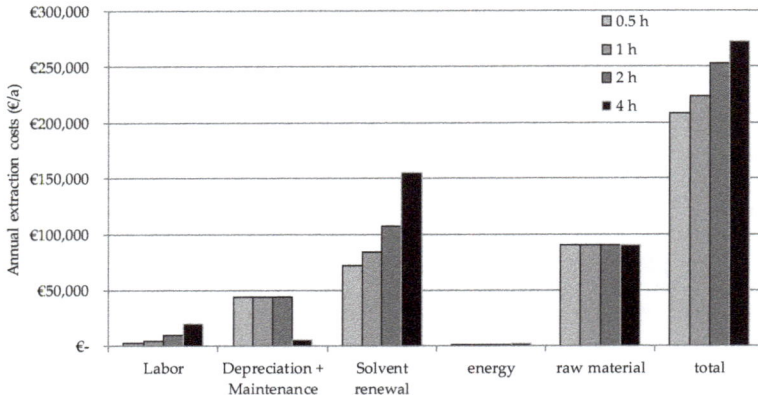

Figure 14. Annual extraction cost reduction by reduced extraction time for Hawthorn.

At three different scales, namely 10 m^3, 5 m^3 and 3 m^3 percolation columns, 2400 t/a of yew needles are extracted. It becomes obvious that the solvent amount can be reduced drastically, as extraction can take place much more rapidly when the particles are ground instead of just chopped.

This also results in lower operation costs, although 20% of the investment costs per year are considered as maintenance expenses compared to only 10% in the benchmark, because a smaller amount of solvent has to be recycled or replaced. The investment costs rise from the benchmark process to the alternative with 10 m^3 percolators, due to the high equipment costs of the mill. The need for smaller apparatus sizes in the 5 m^3 and the 3 m^3 case result in lower investment costs compared to the benchmark. Due to the smaller dimensions of the percolators, the number of batches rises from 375 in the benchmark process to 2500 in the 3 m^3 case. The same occurs with the total process time; that is, roughly 2600 h/a in the benchmark process and 7500 h/a for the 3 m^3 case, which has to be considered as continuously operating. It is obvious that the process proposal provides great potential for:

- reducing the solvent amount,
- minimizing costs for solvent storage, recycling and replacement,
- continuously running fully automated solid-liquid extraction,
- replacing established processes with state-of-the art technology with comparable or even lower CAPEX,
- reducing COG.

Potential for further intensification could be the focus on green extractions technologies [6].

7. Water-Based Green Extraction Processing

Conventional solid-liquid extraction of herbal raws, either for further isolating one single substance or to sell the extract as phytomedicine, often utilizes organic solvents. Despite the advantage of pressurized hot water extraction to extract non-polar substances [46], organic solvents will certainly play an important role over the next decades. Especially for products which are less in demand, solvent costs dominate the whole process, as depicted in Figure 15.

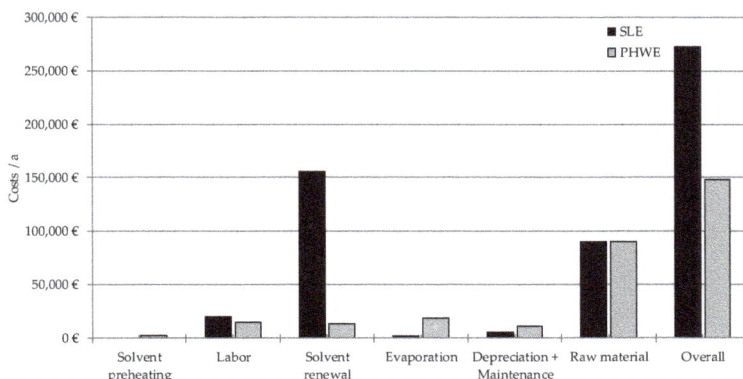

Figure 15. Annual Cost of Goods for hawthorn, Comparison SLE and PHWE [50].

The assumptions for the cost calculation are as follows:

- 30 t of leaves are extracted in 60 batches a year in a multi-purpose herbal raw. The costs for the herbal raw material is 3 €/kg and the yearly capacity of this product is 25%.
- An extraction equipment with 2 m^3 of volume is used. The investment cost is 200,000 €.
- The extraction takes place for four hours and the solvent ratio is 2.7 kg Solvent/kg Herbal raw material/h.
- The extract is evaporated for solvent recycling purpose. Steam is used (120 °C, 5 bar, 2.7 MJ/kg) to operate the evaporator. The costs are 13 €/t which is typical for a site infrastructure.
- 10% of the solvent has to be renewed after each extraction due to loss. Moreover, the whole solvent (20 m^3) is exchanged once a year to maintain a constant product quality. The solvent is priced at 3 €/kg.
- Labor costs are 100,000 €/a.
- The costs for yearly depreciation and maintenance are 2.5% each (multi-purpose herbal raw).

Totally water-based processes exemplify the Green Extraction [6,66,67] approach perfectly. Additionally, they are kosher and halal, directly GRAS, and therefore, represent ideal manufacturing technology for the market demands. A process sequence of PHWE and NF for concentration, followed by purification based on chlorophyll precipitation, liquid-liquid extraction for pre-purification and/or chromatography with final formulation by crystallization or direct lyophilisation seems to be the most direct and logical manufacturing technology approach for the future, efficiently generating reliable product quality under all marked regulation demands.

This could be systematically achieved by the QbD-(Quality by Design) approach, which is demanded by regulatory authorities like FDA and EMA [68–70]. A central part of such innovative approval documentation is manufacturing operation robustness gained by implementation of process analytical technologies (PAT) [44].

8. Inline Process Control in Phyto Extraction

Inline process control is established in the sugar industry for adequate payment of cultivating farmers due to the actual sugar content of beet with the aid of NIR directly on the tractor [71]. NIR is an efficient tool for water content determination [60]. FTIR- and Raman-mapping technologies are state of the art at e.g., JKI Berlin for cultivation origin determination. Moreover, those techniques, as well as CLSM, have proven their usefulness in process design [53,61,63]. Techniques adopted in process development with PCA/PCS-analysis toward the individual component system have the potential to become key-enabling technologies for process control under modern PAT (process analytical technology) approaches within QbD-Process design and operation. Figure 16 exemplifies such a result for the extraction anethol and fenchone.

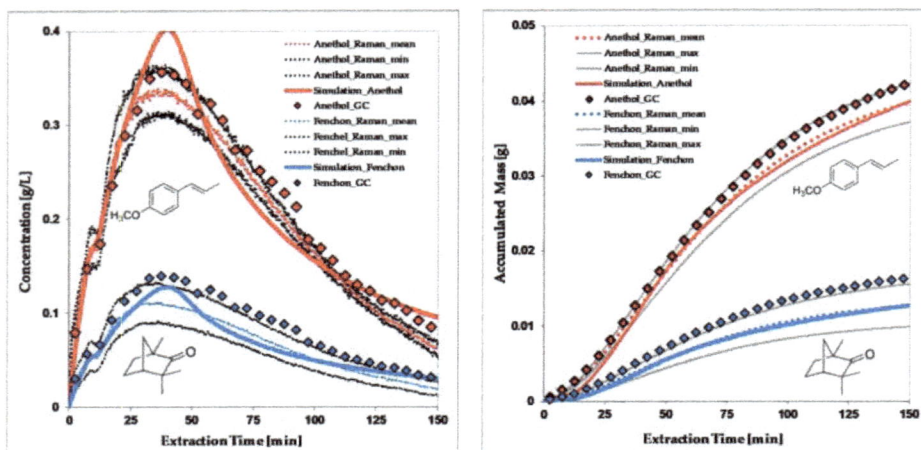

Figure 16. Concentration measurement by raman spectroscopy of anethol (dark grey) and fenchone (light grey) during extraction [44].

9. Options and Opportunities for Future Value Generation

Figure 17 displays a flexible modular manufacturing concept for herbal raw materials at a small scale, which, due to its continuous operation scheduling, may well act as a potential technical solution for SMEs on their way to becoming fully-integrated manufacturers. Only fully-integrated manufacturers with end customer access will be able to gain significant margins, as shown previously. One obstacle for SMEs is investment volume, which needs to be small and refundable for one single product. Here, typical scenarios for either 1 mi. € investment for a standardized extract at 10–50 €/kg dry extract cost or a higher value API at 50–100 €/kg product COGs at 5 mio. € is visualized, e.g., in a container design. In those scenarios, return on investment can be less than 3 years.

Business targets are moving due to changing markets and growing competition, as well as national regulatory decisions. The market demand increases, but industry does not move fast enough to ensure survival. Underlining this point is the fact that most companies do not invest in research and development at all, and are not seeking any product or technology innovation. Not to invest in their own future is not a strategic option. On the contrary, product innovation towards a higher market value is the only protection against increasing regulatory demands.

The market potential of natural products is hampered by efficacy constraints, because they too often lack sufficient definition and testing. QbD, instead of DER, definition is urgently required to cope with natural batch variability in manufacturing operation in order to gain defined products which are within specifications, and with less batch failures.

Manufacturing has not invested sufficiently in recent technology developments over the last 10–15 years of research. Detailed analysis points out that a process of industrialization of research and innovation is needed now to maintain and advance the natural products industry, and has the potential to solve the general challenge of manufacturing cost efficiency and product efficacy in combination with the robustness and reliability issues.

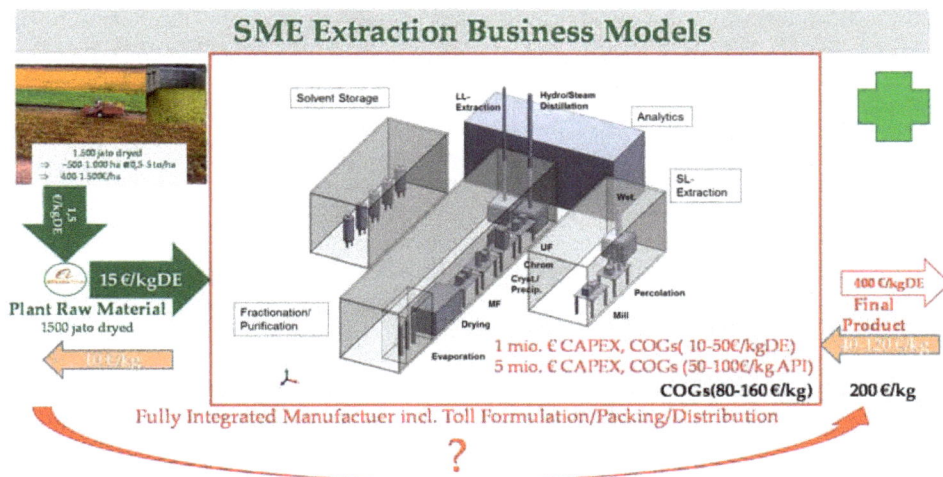

Figure 17. Business model for especially SMEs in phytomedicine manufacturing. CAPEX, capital expenditure; LL, Liquid-Liquid; MF, Microfiltration; UF, Ultrafiltration, SL, Solid-Liquid. (compare Figure 1).

One organizational approach could be the creation of regional agricultural cooperatives, as the sugar and hop industries chose to do, becoming fully integrated manufacturers with moderate investments, thereby gaining the full added value of the end product.

The cost scenarios analyzed show potential for dried extracts (DE) in the 10–50 €/kg range, as well as purified active ingredients (API) in 50–100 €/kg magnitude, which seems realistic within the typical sales price range.

As options for further growth, agricultural co-operatives are needed, alternative state supported infrastructures of technical centers, or medium-size single product manufacturers should be strengthened to enable fast direct access to the technology by first prototype centers. Besides the Clausthal Institute, in Europe, France, seems to be most advanced country for the implemention of such regional strategies, strengthening the local economy from agriculture to pharmaceutical, nutraceutical, aroma/flavors, and cosmetics manufacturing [72].

As a summary, infrastructure and technology exists and is accessible for industrialization, leaving no excuse for decision makers in industry [73].

For example, new active substances are urgently needed against infectious diseases. The development of new antibiotics against resistant microbes is laborious, and obviously does not offer the high commercial incentives expected by big-pharma. Therefore, a clear impulse for industry has to be set to change this situation in favor of research in this field of urgent societal needs by national and international funding organizations, part of which has already been started. Ideas and proposals for either API-substances or herbal sources could, in the meantime, be efficiently and within short timelines transferred to industrialization with the aid of modern process development and design tools, as well as modern manufacturing technologies, as discussed earlier in this contribution. Herbal raw material is still an indisputably innovative source provided by nature, which is far from being fully exploited and industrialized.

10. Conclusions

In conclusion, a suitable approach may be to switch to, or at least to put more emphasis on, standardized extracts, complete with efficacy studies and new approval processes supported by QbD-based process design, which enables process operation at its economical optimum. This is beneficial, if not essential, to maintaining or regaining competitiveness in existing markets. Firstly, the batch variability of typical natural feedstocks can be accounted for in the process design.

Beyond that, it creates the technical basis for addressing increasing societal needs in the product development of innovative, plant-based antibiotics, and/or green and sustainable, resource-efficient manufacturing concepts with additional consumer benefits.

The key role of plants in the medicinal and pharmaceutical fields for thousands of years is undisputed; however, it has had its difficulties. Nevertheless, even today, innovative molecules with therapeutic potential are quite often based on plants, which, in principle, have been known for decades or even centuries. Being able to break down complex natural mixtures into individual molecules or groups of molecules by advanced analytical tools allows the characterization and testing for specific pharmaceutical/medicinal applications. Some of the best examples for successful applications in cancer and malaria treatments are Taxol and Artemisinin.

Maybe the variety of plants and their broad scope of beneficial applications in healthcare might be exploited to find solutions to one of the biggest threats to humankind in this day and age, i.e., the search for new antibiotic effects against MRSA.

Author Contributions: L.U. wrote the paper and contributed to visualization. M.S. developed the process model and did experimental work shown in this review. M.T. gave insight into regulatory aspects. H.S. contributed towards the application of spectroscopic techniques for analysis, specificly the analysis of raw herbal material, H.H. gave additional insight into regulatory aspects and application of herbal products, R.D. contributed towards engineering and economic aspects. J.S. was responsible for conception and supervision.

Funding: The authors want to thank the Bundesministerium für Wirtschaft und Energie (BMWi), especially M. Gahr (Projektträger FZ Jülich), for funding this scientific work. We also acknowledge the financial support obtained from the Deutsche Forschungsgemeinschaft (DFG) in Bonn, Germany (project Str 586/4-2).

Acknowledgments: The authors like to thank their colleagues from ProcessNet Working Group on "Phyto-Extraktion-Products and Processes" for challenging discussions concerning status and trends of natural products extraction. Gracious thanks are extended to the experienced and motivated laboratory team of the Clausthal Institute. Moreover, all the lectures of the annual education and training courses, perfectly organized by FAH Bonn and PDA Berlin, in Clausthal are kindly acknowledged for their efforts and input.

Conflicts of Interest: The authors declare no conflict of interest.

Abbreviations

a_F	Particle surface, m^2
CAPEX	Capital expenitures
c_L	Concentration in the liquid phase, kg/m^3
COG	Cost of goods
c_P	Concentration in the porous particle, kg/m^3
D_{ax}	Axial dispersion coefficient, m/s^2
DE	dry extract
D_{eff}	Effective diffusion coefficient, m^2/s
DER	drug extract ratio
d_F	Particle diameter m
DPF	Distributed plug flow
EMPL	employee
K_L	Equilibrium constant, m^3/kg
k_f	Mass transport coefficient, m/s
Pe	Péclet number
PSD	Particle size distribution
q	Loading, kg/m^3
q_{max}	Maximum Loading, kg/m^3
Re	Reynolds number
r	Radius, m
Sc	Schmidt number
Sh	Sherwood number

SME	Small and medium-sized enterprise
SP	selling price
t	Time, s
u_z	Superficial velocity, m/s
V	Volume flow, m^3/s
z	Coordinate in axial direction, m
ε	Voids fraction, -
ρ	Density, kg/m^3

References

1. Cragg, G.M.; Newman, D.J. Natural products: A continuing source of novel drug leads. *Biochim. Biophys. Acta* **2013**, *1830*, 3670–3695. [CrossRef] [PubMed]

2. Li, W.L.; Zheng, H.C.; Bukuru, J.; de Kimpe, N. Natural medicines used in the traditional Chinese medical system for therapy of diabetes mellitus. *J. Ethnopharmacol.* **2004**, *92*, 1–21. [CrossRef] [PubMed]

3. Atanasov, A.G.; Waltenberger, B.; Pferschy-Wenzig, E.-M.; Linder, T.; Wawrosch, C.; Uhrin, P.; Temml, V.; Wang, L.; Schwaiger, S.; Heiss, E.H.; et al. Discovery and resupply of pharmacologically active plant-derived natural products: A review. *Biotechnol. Adv.* **2015**, *33*, 1582–1614. [CrossRef] [PubMed]

4. Ditz, R.; Gerard, D.; Hagels, H.; Igl, N.; Schäffler, M.; Schulz, H.; Stürtz, M.; Tegtmeier, M.; Treutwein, J.; Strube, J.; et al. *Phytoextracts: Proposal towards a New Comprehensive Research Focus*; DECHEMA Gesellschaft für Chemische Technik und Biotechnologie e.V.: Frankfurt, Germany, 2017.

5. Bart, H.-J.; Pilz, S. *Industrial Scale Natural Products Extraction*; Wiley-VCH: Weinheim, Germany, 2011.

6. Chémat, F.; Strube, J. (Eds.) *Green Extraction of Natural Products: Theory and Practice*; Wiley VCH: Weinheim, Germany, 2015.

7. Veeresham, C. Natural products derived from plants as a source of drugs. *J. Adv. Pharm. Technol. Res.* **2012**, *3*, 200–201. [CrossRef] [PubMed]

8. Newman, D.J.; Cragg, G.M. Natural products as sources of new drugs over the last 25 years. *J. Nat. Prod.* **2007**, *70*, 461–477. [CrossRef] [PubMed]

9. Lopez, S.N.; Ramallo, I.A.; Sierra, M.G.; Zacchino, S.A.; Furlan, R.L.E. Chemically engineered extracts as an alternative source of bioactive natural product-like compounds. *Proc. Natl. Acad. Sci. USA* **2007**, *104*, 441–444. [CrossRef] [PubMed]

10. Mahmood, H.Y.; Jamshidi, S.; Sutton, J.M.; Rahman, K.M. Current Advances in Developing Inhibitors of Bacterial Multidrug Efflux Pumps. *Curr. Med. Chem.* **2016**, *23*, 1062–1081. [CrossRef] [PubMed]

11. Posadzki, P.; Watson, L.; Ernst, E. Herb-drug interactions: An overview of systematic reviews. *Br. J. Clin. Pharmacol.* **2013**, *75*, 603–618. [CrossRef] [PubMed]

12. Newman, D. Screening and identification of novel biologically active natural compounds. *F1000Research* **2017**, *6*, 783. [CrossRef] [PubMed]

13. Ramallo, I.A.; Salazar, M.O.; Mendez, L.; Furlan, R.L.E. Chemically Engineered Extracts: Source of Bioactive Compounds. *Acc. Chem. Res.* **2011**, *44*, 241–250. [CrossRef] [PubMed]

14. Ta, C.; Arnason, J. Mini Review of Phytochemicals and Plant Taxa with Activity as Microbial Biofilm and Quorum Sensing Inhibitors. *Molecules* **2016**, *21*, 29. [CrossRef] [PubMed]

15. Kalia, V.C.; Wood, T.K.; Kumar, P. Evolution of Resistance to Quorum-Sensing Inhibitors. *Microb. Ecol.* **2014**, *68*, 13–23. [CrossRef] [PubMed]

16. Borges, A.; Abreu, A.; Dias, C.; Saavedra, M.; Borges, F.; Simões, M. New Perspectives on the Use of Phytochemicals as an Emergent Strategy to Control Bacterial Infections Including Biofilms. *Molecules* **2016**, *21*, 877. [CrossRef] [PubMed]

17. Aparna, V.; Dineshkumar, K.; Mohanalakshmi, N.; Velmurugan, D.; Hopper, W. Identification of Natural Compound Inhibitors for Multidrug Efflux Pumps of Escherichia coli and Pseudomonas aeruginosa Using In Silico High-Throughput Virtual Screening and In Vitro Validation. *PLoS ONE* **2014**, *9*, e101840. [CrossRef] [PubMed]

18. Nascimento, G.G.F.; Locatelli, J.; Freitas, P.C.; Silva, G.L. Antibacterial activity of plant extracts and phytochemicals on antibiotic-resistant bacteria. *Braz. J. Microbiol.* **2000**, *31*, 247–256. [CrossRef]

19. Efferth, T.; Kahl, S.; Paulus, K.; Adams, M.; Rauh, R.; Boechzelt, H.; Hao, X.; Kaina, B.; Bauer, R. Phytochemistry and pharmacogenomics of natural products derived from traditional Chinese medicine and Chinese materia medica with activity against tumor cells. *Mol. Cancer Ther.* **2008**, *7*, 152–161. [CrossRef] [PubMed]

20. Europäisches Zentrum für die Prävention und die Kontrolle von Krankheiten. *The Bacterial Challenge, Time to React: A Call to Narrow the Gap between Multidrug-Resistant Bacteria in the EU and the Development of New Antibacterial Agents*; ECDC: Stockholm, Sweden, 2009.

21. DECHEMA e.V. Biobased World. Available online: https://www.biobasedworldnews.com/ (accessed on 22 June 2018).

22. Bundesministerium für Bildung und Forschung (BMBF). *Wegweiser Bioökonomie—Forschung für Biobasiertes und Nachhaltiges Wirtschaftswachstum*; BMBF: Berlin, Germany, 2014.

23. Jamshidi Aidji, M.; Morlock, G. *Effect Directed Analysis of Salvia Officinalis, Poster Presentation*; University of Giessen: Gießen, Germany, 2014.

24. Morlock, G.; Schwack, W. Hyphenations in planar chromatography. *J. Chromatogr. A* **2010**, *1217*, 6600–6609. [CrossRef] [PubMed]

25. Staniek, A.; Bouwmeester, H.; Fraser, P.D.; Kayser, O.; Martens, S.; Tissier, A.; van der Krol, S.; Wessjohann, L.; Warzecha, H. Natural products—Learning chemistry from plants. *Biotechnol. J.* **2014**, *9*, 326–336. [CrossRef] [PubMed]

26. European Medicines Agency. *Assessment Report on Echinacea purpurea (L.) Moench, Radix*; University of Ljubljana: Ljubljana, Slovenia, 2017.

27. European Medicines Agency. *European Union Herbal Monograph on Echinacea purpurea (L.) Moench, Radix*; European Medicines Agency: London, UK, 2017.

28. Council of Europe. *European Pharmacopoeia*, 8th ed.; Published in Accordance with the Convention on the Elaboration of a European Pharmacopoeia (European Treaty Series No. 50); Council of Europe: Strasbourg, France, 2013–2015.

29. Food and Drug Administration (FDA). *Guidance for Industry—Botanical Drug Development*; FDA: Silver Spring, MD, USA, 2016.

30. Food and Drug Administration (FDA). Dietary Supplements. Available online: https://www.fda.gov/Food/DietarySupplements/ProductsIngredients/default.htm (accessed on 21 September 2018).

31. Chinese Food and Drug Administration (CFDA). Verification and Issuance of Registration Certificates for Imported (Incl. from Hong Kong, Macao and Taiwan) TCM and Natural Medicine. Available online: http://eng.sfda.gov.cn/WS03/CL0769/98141.html (accessed on 21 September 2018).

32. Akerele, O. WHO traditional medicines programme: Progress and perspectives. *WHO Chron.* **1984**, *38*, 76–81. [PubMed]

33. Chaing, H.S.; Merino-Chavez, G.; Yang, L.L.; Hafez, E.S.E. Medical plants: Conception/Contratception. *Adv. Contracept Deliv Syst.* **1994**, *10*, 355–363. [PubMed]

34. Chotchoungchatchai, S.; Saralamp, P.; Jenjittikul, T.; Pornsiripongse, S.; Prathanturarug, S. Medicinal plants used with Thai Traditional Medicine in modern healthcare services: A case study in Kabchoeng Hospital, Surin Province, Thailand. *J. Ethnopharmacol.* **2012**, *141*, 193–205. [CrossRef] [PubMed]

35. Wagner, H.; Ulrich-Merzenich, G. (Eds.) *Evidence and Rational Based Research on Chinese Drugs*; Springer: Vienna, Austria, 2013.

36. Giovannini, P.; Howes, M.-J.R.; Edwards, S.E. Medicinal plants used in the traditional management of diabetes and its sequelae in Central America: A review. *J. Ethnopharmacol.* **2016**, *184*, 58–71. [CrossRef] [PubMed]

37. Govindaraghavan, S.; Sucher, N.J. Quality assessment of medicinal herbs and their extracts: Criteria and prerequisites for consistent safety and efficacy of herbal medicines. *Epilepsy Behav.* **2015**, *52*, 363–371. [CrossRef] [PubMed]

38. Pferschy-Wenzig, E.-M.; Bauer, R. The relevance of pharmacognosy in pharmacological research on herbal medicinal products. *Epilepsy Behav.* **2015**, *52*, 344–362. [CrossRef] [PubMed]

39. Russo, G.L. Ins and outs of dietary phytochemicals in cancer chemoprevention. *Biochem. Pharmacol.* **2007**, *74*, 533–544. [CrossRef] [PubMed]

40. Schuster, B.G. A new integrated program for natural product development. *J. Altern. Complement. Med.* **2001**, *7*, 61–72. [CrossRef]

41. Food and Drug Administration (FDA). *Guidance for Industry. PAT—A Framework for Innovative Pharmaceutical Development, Manufacturing, and Quality Assurance*; FDA: Silver Spring, MD, USA, 2004.

42. Food and Drug Administration (FDA). *Pharmaceutical cGMP for the 21st Century: A Risk Based Approach*; FDA: Silver Spring, MD, USA, 2004.

43. Sixt, M.; Koudous, I.; Strube, J. Process design for integration of extraction, purification and formulation with alternative solvent concepts. *C. R. Chim.* **2016**, *19*, 733–748. [CrossRef]

44. Sixt, M.; Gudi, G.; Schulz, H.; Strube, J. In-line Raman spectroscopy and advanced process control for the extraction of anethole and fenchone from fennel (*Foeniculum vulgare* L. MILL.). *C. R. Chim.* **2018**, *21*, 97–103. [CrossRef]

45. Sixt, M.; Uhlenbrock, L.; Strube, J. Toward a Distinct and Quantitative Validation Method for Predictive Process Modelling—On the Example of Solid-Liquid Extraction Processes of Complex Plant Extracts. *Processes* **2018**, *6*, 66. [CrossRef]

46. Sixt, M.; Strube, J. Pressurized hot water extraction of 10-deacetylbaccatin III from yew for industrial application. *Resour. Effic. Technol.* **2017**, *3*, 177–186. [CrossRef]

47. Uhlenbrock, L.; Sixt, M.; Strube, J. Quality-by-Design (QbD) process evaluation for phytopharmaceuticals on the example of 10-deacetylbaccatin III from yew. *Resour. Effic. Technol.* **2017**, *3*, 137–143. [CrossRef]

48. Wübbeke, J.; Meissner, M.; Zenglein, M.J.; Ives, J.; Conrad, B. *Made in China 2025: The Making of a High-Tech Superpower and Consequences for Industrial Countries*; Mercator Institute for China Studies: Berlin, Germany, 2016.

49. Sixt, M.; Strube, J. Systematic Design and Evaluation of an Extraction Process for Traditionally Used Herbal Medicine on the Example of Hawthorn (*Crataegus monogyna* JACQ.). *Processes* **2018**, *6*, 73. [CrossRef]

50. Sixt, M. *Entwicklung von Methoden zur systematischen Gesamtprozessentwicklung und Prozessintensivierung von Extraktions- und Trennprozessen zur Gewinnung pflanzlicher Wertkomponenten*; Clausthal University of Technology: Clausthal-Zellerfeld, Germany; Shaker: Aachen, Germany, 2018.

51. Kassing, M.; Jenelten, U.; Schenk, J.; Hänsch, R.; Strube, J. Combination of Rigorous and Statistical Modeling for Process Development of Plant-Based Extractions Based on Mass Balances and Biological Aspects. *Chem. Eng. Technol.* **2012**, *35*, 109–132. [CrossRef]

52. Kassing, M.; Jenelten, U.; Schenk, J.; Strube, J. A New Approach for Process Development of Plant-Based Extraction Processes. *Chem. Eng. Technol.* **2010**, *33*, 377–387. [CrossRef]

53. Kaßing, M. *Process Development for Plant-Based Extract Production*; Clausthal University of Technology: Clausthal-Zellerfeld, Germany; Shaker: Aachen, Germany, 2012.

54. Both, S.; Eggersglüß, J.; Lehnberger, A.; Schulz, T.; Schulze, T.; Strube, J. Optimizing Established Processes like Sugar Extraction from Sugar Beets—Design of Experiments versus Physicochemical Modeling. *Chem. Eng. Technol.* **2013**, *36*, 2125–2136. [CrossRef]

55. Both, S. *Systematische Verfahrensentwicklung für Pflanzlich Basierte Produkte im Regulatorischen Umfeld*; Clausthal University of Technology: Clausthal-Zellerfeld, Germany; Shaker: Aachen, Germany, 2015.

56. Both, S.; Chemat, F.; Strube, J. Extraction of polyphenols from black tea—Conventional and ultrasound assisted extraction. *Ultrason Sonochem.* **2014**, *21*, 1030–1034. [CrossRef] [PubMed]

57. Ndocko Ndocko, E.; Bäcker, W.; Strube, J. Process Design Method for Manufacturing of Natural Compounds and Related Molecules. *Sep. Sci. Technol.* **2008**, *43*, 642–670. [CrossRef]

58. Goedecke, R. *Fluidverfahrenstechnik*; Wiley-VCH: Hoboken, NI, USA, 2008.

59. Sixt, M.; Strube, J. Systematic and Model-Assisted Evaluation of Solvent Based- or Pressurized Hot Water Extraction for the Extraction of Artemisinin from *Artemisia annua* L. *Processes* **2017**, *5*, 86. [CrossRef]

60. Koudous, I. *Stoffdatenbasierte Verfahrensentwicklung zur Isolierung von Wertstoffen aus Pflanzenextrakten*; Clausthal University of Technology: Clausthal-Zellerfeld, Germany; Shaker: Aachen, Germany, 2016.

61. Gudi, G.; Krähmer, A.; Koudous, I.; Strube, J.; Schulz, H. Infrared and Raman spectroscopic methods for characterization of *Taxus baccata* L.—Improved taxane isolation by accelerated quality control and process surveillance. *Talanta* **2015**, *143*, 42–49. [CrossRef] [PubMed]

62. Duke, M.V.; Paul, R.N.; Elsohly, H.N.; Sturtz, G.; Duke, S.O. Localization of Artemisinin and Artemisitene in Foliar Tissues of Glanded and Glandless Biotypes of *Artemisia annua* L. *Int. J. Plant. Sci* **1994**, *155*, 365–372. [CrossRef]

63. Gudi, G.; Krähmer, A.; Krüger, H.; Hennig, L.; Schulz, H. Discrimination of Fennel Chemotypes Applying IR and Raman Spectroscopy: Discovery of a New γ-Asarone Chemotype. *J. Agric. Food Chem.* **2014**, *62*, 3537–3547. [CrossRef] [PubMed]

64. Lucke, M.; Koudous, I.; Sixt, M.; Huter, M.J.; Strube, J. Integrating crystallization with experimental model parameter determination and modeling into conceptual process design for the purification of complex feed mixtures. *Chem. Eng. Res. Des.* **2018**, *133*, 264–280. [CrossRef]

65. Helling, C.; Strube, J. *Modeling and Experimental Model Parameter Determination with Quality by Design for Bioprocesses: Biopharmaceutical Production Technology*; Wiley-VCH-Verl: Weinheim, Germany, 2012; pp. 409–445.

66. Chemat, F.; Vian, M.A. (Eds.) *Alternative Solvents for Natural Products Extraction*; Springer: Berlin/Heidelberg, Germany, 2014.

67. Chémat, F.; Vorobiev, E.; Lebovka, N.I. *Enhancing Extraction Processes in the Food Industry*; CRC Press: Boca Raton, FL, USA, 2012.

68. European Medicines Agency, Food and Drug Administration. *EMA-FDA Pilot Program for Parallel Assessment of Quality by Design Applications*; European Medicines Agency, Food and Drug Administration: London, UK, 2011.

69. Food and Drug Administration (FDA). *Guideline for Implimentation of Q9*; FDA: Silver Spring, MD, USA, 2006.

70. ICH Expert Working Group (ICH EWG). *Riskmanagement (Q9). ICH Harmonised Tripartite Guideline. Q9*; ICH Expert Working Group: Geneva, Switzerland, 2005.

71. Roggo, Y.; Duponchel, L.; Huvenne, J.-P. Quality Evaluation of Sugar Beet (*Beta vulgaris*) by Near-Infrared Spectroscopy. *J. Agric. Food Chem.* **2004**, *52*, 1055–1061. [CrossRef] [PubMed]

72. Chémat, F.; Fernandez, X. (Eds.) *La Chimie des Huiles Essentielles: Tradition et Innovation*; Vuibert: Paris, France, 2012.

73. Forschungsvereinigung der Arzneimittel-Hersteller e.V. (FAH). *(FAH). Trainings-Course on Phytoextraction—Process Design and Operation, Experiments and Modelling October 2017 at the Institute for Separation and Process Technology*; Clausthal University of Technology: Clausthal-Zellerfeld, Germany, 2017.

MDPI

St. Alban-Anlage 66

4052 Basel

Switzerland

Tel. +41 61 683 77 34

Fax +41 61 302 89 18

www.mdpi.com

Processes Editorial Office

E-mail: processes@mdpi.com

www.mdpi.com/journal/processes

www.ingramcontent.com/pod-product-compliance
Lightning Source LLC
Chambersburg PA
CBHW051855210326
41597CB00033B/5903